MULTIFRACTALS

THEORY and APPLICATIONS

DAVID HARTE

CRC Press
Taylor & Francis Group
Boca Raton London New York

CRC Press is an imprint of the
Taylor & Francis Group, an **informa** business

A CHAPMAN & HALL BOOK

First published 2001 by Chapman & Hall

Published 2019 by CRC Press
Taylor & Francis Group
6000 Broken Sound Parkway NW, Suite 300
Boca Raton, FL 33487-2742

© 2001 by Taylor & Francis Group, LLC
CRC Press is an imprint of Taylor & Francis Group, an Informa business

First issued in paperback 2019

No claim to original U.S. Government works

ISBN-13: 978-0-367-45520-0 (pbk)
ISBN-13: 978-1-58488-154-4 (hbk)

Visit the Taylor & Francis Web site at
http://www.taylorandfrancis.com

and the CRC Press Web site at
http://www.crcpress.com

Library of Congress Cataloging-in-Publication Data

Harte, David.
    Multifractals : theory and applications / David Harte.
        p. cm.
    Includes bibliographical references and index.
    ISBN 1-58488-154-2
    1. Multifractals. I. Title.

QA614.86 .H35 2001
514′.742—dc21

2001028886

Library of Congress Card Number 2001028886

MULTIFRACTALS

THEORY and APPLICATIONS

Preface

Multifractal theory is essentially rooted in probability theory, though draws on complex ideas from each of physics, mathematics, probability theory and statistics. It has also been used in a wide range of application areas: dynamical systems, turbulence, rainfall modelling, spatial distribution of earthquakes and insect populations, financial time series modelling and internet traffic modelling.

I have approached the subject as a statistician and applied probabilist, being interested initially in calculating fractal dimensions of spatial point patterns produced by earthquakes. Since the subject of multifractals draws on theory from a number of disciplines and also has applications in a number of different areas, there is an inevitable difficulty arising from different terminology, concepts, and levels of technical rigor. I have attempted to pull together ideas from all of these areas and place the material into a probabilistic and statistical context, using a language that makes them accessible and useful to statistical scientists. It was my intention, in particular, to provide a framework for the evaluation of statistical properties of estimates of the Rényi fractal dimensions.

It should not be interpreted that the book is only of interest to statisticians. The estimation of fractal dimensions from a statistical perspective is virtually uncovered in other books. We attempt to categorise forms of bias as intrinsic or extrinsic and describe their effect on the dimension estimates. Intrinsic biases are those effects which are caused by an inherent characteristic of the probability distribution, whereas extrinsic bias refers to those characteristics that are caused by sampling and other methodological difficulties. Examples of such biases are given using known mathematical and statistical models.

The main emphasis in the book is on multifractal *measures*. More recent developments on stochastic processes that are multiscaling and sometimes referred to as 'multifractal' stochastic processes, compared to those self-similar stochastic processes that are monoscaling, are peripheral to the main direction of material contained in this book. These 'multifractal' stochastic processes are only mentioned briefly.

The first part of the book provides introductory material and different definitions of a multifractal measure, in particular, those constructions based on lattice coverings and point-centred coverings by spheres. In the second part, it is shown that the so called 'multifractal formalism' for these two constructions can be justified using a standard probabilistic technique, namely the theory of large deviations. The final part presents estimators of Rényi dimensions, of integer order two and greater, and discusses their properties. It also discusses various applications of dimension estimation, and provides a detailed case study of spatial point patterns of earthquake locations. A brief summary of some definitions of dimension, and results from the theory of large deviations, is included in the Appendices.

One cannot hope to wrap up all of the information required by the statistician in one book. Indeed that is not my intention. The Appendices include various definitions of the dimension of a set, and some results from the theory of large deviations. However, when further information is required, there are other very substantial works that should be consulted. I will mention a few in particular. The book by Falconer (1990) gives an excellent overview of fractals from a geometrical perspective. He gives a thorough treatment of the various definitions of dimension and their relationships. The later book by Falconer (1997) contains much new material and various techniques that are of use in studying the mathematics of fractals with more emphasis on measures. Ruelle (1989) gives a nice introduction to dynamical systems, which have provided much of the motivation for the study of multifractal theory. Ellis (1985) provides a detailed account of the theory of large deviations, and the excellent and very readable book by Abarbanel (1995) deals with non-linear data analysis from more of a physics perspective.

I have closely followed some definitions, statements of theorems, and other text contained in the publications listed below. Particular subsections of this book for which copyright permissions were requested are listed: from Mandelbrot (1989), *PAGEOPH*, quoted text in §1.8; Cutler (1991), *Journal of Statistical Physics*, §2.4.3, §2.7.5, §2.7.7; Holley & Waymire (1992), *Annals of Applied Probability*, §6.3.2, §6.3.3, §6.3.4, §6.3.5, §6.3.11, §6.3.12, §6.3.14; Cutler (1997), *Fields Institute Communications*, §10.5.1(2,3), §10.5.2, §10.5.3, §10.5.4; Falconer (1990), Wiley, Chichester, various marked extracts in Appendix A; Ellis (1984), *Annals of Probability*, §B.3.8, §B.3.14, §B.3.17; and Ellis (1985), Springer-Verlag, New York, §B.3.1, §B.3.4, §B.3.5, §B.3.6, §B.3.11, §B.3.15. I would like to thank the above authors and also others whose work I have referred to in the book. The following figures have been adapted, with permission, from Harte (1998), *Journal of Nonlinear Science*: 1.6, 1.8, 1.7, 9.3, 11.4, 11.2, 11.1 and 11.3.

This book had its beginnings in a reading group at the Victoria University of Wellington in 1993. Members of the group were Professor David Vere-Jones, and Drs. Robert Davies, Thomas Mikosch and Qiang Wang. We were interested in the estimation and interpretation of 'fractal' dimensions, in particular, in the earthquake and meteorological application areas. I would like to thank all members of the group for their help, for the many interesting hours we had trying to interpret various dimension plots, and their continued interest since then. I would also like to thank Peter Thomson for his encouragement and interest in the project, and would particularly like to thank David Vere-Jones for his help and continued encouragement over a number of years.

David Harte
May 2001

List of Figures

List of Notation

Contents

PART I

INTRODUCTION AND PRELIMINARIES

CHAPTER 1

Motivation and Background

1.1 Introduction

An intuitive introduction is given in this chapter to a number of the main concepts that will be discussed in the following chapters of the book. In §1.2 we describe the difference between a *fractal set* and a *multifractal measure*. Fractal and multifractal methods have been used extensively in the description of dynamical systems which are introduced in §1.3. A distinguishing feature of processes that have multifractal characteristics is that various associated probability distributions display powerlaw properties. Other application areas where powerlaw scaling characteristics have been discussed extensively in the literature are in the fields of turbulence, rainfall, and earthquake modelling. These are discussed in §1.4, §1.5, and §1.6 respectively.

The character of this chapter is intentionally different from that of subsequent chapters. The emphasis in this chapter is on a descriptive introduction, and often terminology, in particular, 'dimension' and 'fractal' will be used loosely. Formal definitions of most concepts will be given in subsequent chapters. Definitions of various dimensions of a set, and some of their inter relationships, can be found in Appendix A. A detailed and very elegant account can be found in the book by Falconer (1990).

1.2 Fractal Sets and Multifractal Measures

The books by Mandelbrot (1977, 1983) have initiated considerable interest in describing objects with an extremely irregular shape. His examples included galaxies, lengths of coastlines, snowflakes, and the Cantor set. Some of these objects have, what appear at least initially, some rather bizarre characteristics, for example, coastlines of infinite length and snowflakes with an infinite surface area. This tends to happen when the set is very irregular, and further, characteristics of the set at a given level of magnification are essentially the same as those at other levels of magnification apart from a scale factor; hence irregularity is repeated on finer and finer levels ad infinitum. These sets are referred to as being self-similar (see Appendix §A.1).

One way to describe the size of these sets is to calculate its 'fractal' dimension. For example, the dimension of an irregular coastline may be greater than one but less than two, indicating that it is not like a simple line and has space filling characteristics in the plane. Likewise, the surface area of a snowflake may be greater than two but less than three, indicating that its surface is more complex than regular geometrical shapes, and is partially volume filling. A definition of what is a fractal and hence 'fractal' dimension, is still not generally accepted. Intuitively, a fractal set is an object that is extremely irregular, and has a 'dimension' that is fractional. However, we are not so interested in fractal sets per se, but more in measures supported on such sets. Often these measures will be probability measures. This may be made clearer in the following example.

Earthquake occurrence may also have fractal like characteristics. It is thought that earthquakes occur on faults which are essentially a fracture in the earth's crust. A simple clean cut in a three-dimensional object would have a dimension of two. However, consider the situation where small faults branch off larger faults, and from these smaller faults, even smaller faults are found. And this replication is repeated many times to a finer and finer level. If this hypothesis were true, then one would expect the dimension to be greater than two but less than three; i.e., in the vicinity of a large fault, the fracturing would have some volume filling characteristics. However, within that fault network, there are certain areas that will be much more active than others; i.e., have a greater probability of an earthquake event. As such, we could think of the set of possible locations of where an earthquake could occur to be a fractal set, but on that fractal set is a probability measure which describes the likelihood of an event. Usually this probability distribution is extremely irregular to the extent that it does not have a density. The question of interest is: how does one characterise and describe such a probability distribution? This is one of the underlying themes of this book.

Examples of fractal sets with an associated probability measure can be easily constructed. One of the most simple examples is the Cantor measure.

1.2.1 Example - Cantor Measure

The Cantor set is constructed by removing the middle third from the unit interval, then the remaining two subintervals have their middle thirds removed, and this continues ad infinitum, that is

$$
\begin{aligned}
\mathcal{K}_0 &= [0,1] \\
\mathcal{K}_1 &= [0,\tfrac{1}{3}] \cup [\tfrac{2}{3},1] \\
\mathcal{K}_2 &= [0,\tfrac{1}{9}] \cup [\tfrac{2}{9},\tfrac{1}{3}] \cup [\tfrac{2}{3},\tfrac{7}{9}] \cup [\tfrac{8}{9},1] ,
\end{aligned}
$$

etc. The Cantor set is then defined as $\mathcal{K} = \bigcap_{n=0}^{\infty} \mathcal{K}_n$. Note that $\mathcal{K}_n$ contains two scaled copies of $\mathcal{K}_{n-1}$, i.e.,

$$
\mathcal{K}_n = \left(\frac{\mathcal{K}_{n-1}}{3}\right) \bigcup \left(\frac{2}{3} + \frac{\mathcal{K}_{n-1}}{3}\right) \qquad n = 0, 1, \cdots .
$$

It can be seen that the Cantor set contains all numbers in the unit interval whose base 3 expansion does not contain the digit 1.

How do we describe the size of the Cantor set? It can be seen that the Lebesgue measure of $\mathcal{K}_n$ is $\left(\frac{2}{3}\right)^n \to 0$ as $n \to \infty$. Another way is to calculate its dimension. The basic idea of a dimension, d_0, is that it relates to the number of covers that are required to cover the set of interest. For example, 2 boxes of width $\frac{1}{3}$ are required to cover $\mathcal{K}_1$, 4 boxes of width $\frac{1}{9}$ are required to cover $\mathcal{K}_2$, etc. That is, let $N_\delta(\mathcal{K})$ be the number of boxes of width δ that are required to cover the set $\mathcal{K}$, then

$$\frac{\log N_{\delta_n}(\mathcal{K})}{-\log \delta_n} = \frac{\log(2^n)}{-\log(3^{-n})} = \log_3 2,$$

where $\delta_n = 3^{-n}$. The number $\log_3 2$ is the dimension (both box and Hausdorff) of the Cantor set. It can be seen that the required number of boxes scales with the dimension, i.e., $N_{\delta_n}(\mathcal{K}) \sim \delta_n^{-d_0}$ as $n \to \infty$.

The above description of dimension relates more closely to the box counting dimension, though not exactly (see Definition A.3.1). There are many definitions of dimension, some differences being whether the covers are disjoint or overlapping, boxes or spheres, of a fixed width or variable width no greater than δ, and the manner in which the limit $\delta \to \infty$ is taken; however, the basic idea of counts of covers is the same. A summary can be found in Appendix A, and a fuller treatment can be found in Falconer (1990).

Now extend the example further by allocating a mass or probability to each subinterval at each division. In this example, we will allocate $\frac{2}{3}$ of the existing probability in an interval being divided to the right-hand subinterval, and $\frac{1}{3}$ to the left as in Figure 1.1. By construction, the Cantor set is closed and is therefore the support of this measure. Hence the dimension of the support is $\log_3 2$. However, it can be seen that this dimension would be the same regardless of how one allocated

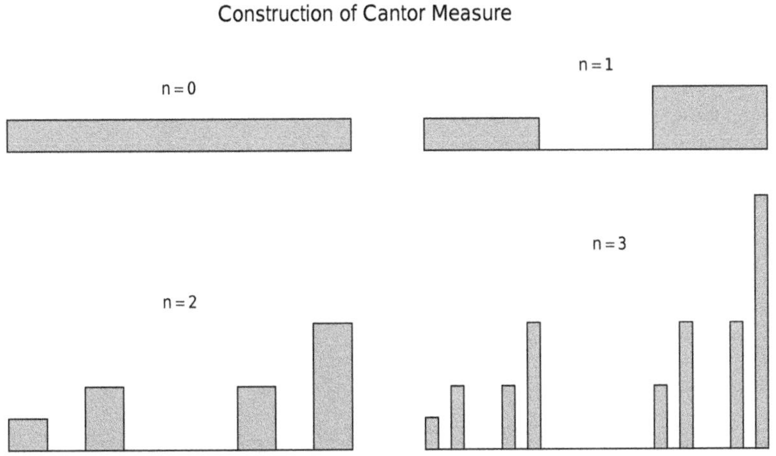

Construction of Cantor Measure

Figure 1.1 *Construction of the Cantor measure with $\frac{1}{3}$ of the probability allocated to the left subinterval and $\frac{2}{3}$ allocated to the right.*

the probabilities. Therefore, how does one describe this probability distribution supported by the Cantor set? Clearly the distribution does not have a density as the Lebesgue measure of the set is zero.

Consider quantifying the rate of probability change in the first bar of $\mathcal{K}_n$ for $n = 1, 2, \cdots$ as follows. The width of this bar is $\delta_n = 3^{-n}$. Let the probability measure at the nth step be denoted by μ_n, then it can be seen that, for all n,

$$\frac{\log \mu_n([0, 3^{-n}])}{\log \delta_n} = \frac{\log(3^{-n})}{\log(3^{-n})} = 1.$$

Similarly, the last bar can be characterised, for all n, as

$$\frac{\log \mu_n([1 - 3^{-n}, 1])}{\log \delta_n} = \frac{\log((2/3)^n)}{\log(3^{-n})} = 1 - \log_3 2 \approx 0.3691.$$

Now we generalise the above description to all subintervals. Let $K_n(y)$ be the set containing subintervals of width δ_n such that if $J \in K_n(y)$, then

$$\frac{\log \mu_n(J)}{\log \delta_n} = y.$$

Also let $\#K_n(y)$ be the number of subintervals of length δ_n contained in $K_n(y)$, then it can be seen from Figure 1.1 that when $n = 3$, $\#K_3(0.3691) = 1$,

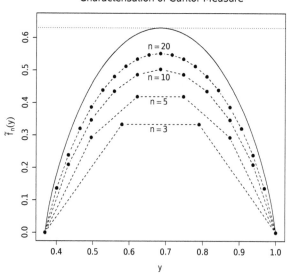

Figure 1.2 *Characterisation of the Cantor measure with $\frac{1}{3}$ of the probability allocated to the left subinterval and $\frac{2}{3}$ allocated to the right. The points are $\widetilde{f}_n(y)$ for $n = 3, 5, 10, 20$, and the solid line is $\lim_{n \to \infty} \widetilde{f}_n(y)$. The dotted line marks the dimension of the Cantor set, i.e., $\log_3 2$.*

$\#K_3(0.5794) = 3$, $\#K_3(0.7897) = 3$, and $\#K_3(1) = 1$. Thus the number of boxes of width $\delta_3 = 3^{-3}$ that are required to cover, for example, $K_3(0.7897)$ is three. Now consider only those values of y where $\#K_n(y) > 0$, where we define $\widetilde{f}_n(y)$ as

$$\widetilde{f}_n(y) = \frac{\log \#K_n(y)}{\log \delta_n}.$$

The function $\widetilde{f}_n(y)$ is plotted in Figure 1.2 for $n = 3, 5, 10$, and 20. The function $\widetilde{f}(y) = \lim_{n\to\infty} \widetilde{f}_n(y)$ is called the *multifractal spectrum*, and is also plotted in Figure 1.2. It can be seen that if the measure was allocated evenly at each division, i.e., $\frac{1}{2}$ of the probability to the left and right, then the multifractal spectrum would exist at only one point, i.e., $\widetilde{f}(\log_3 2) = \log_3 2$. This does not coincide with the maximum of the graph in Figure 1.2, and the difference between the two forms of $\widetilde{f}(y)$ reflects the difference between the two limiting probability distributions. There are a number of interpretations of $\widetilde{f}(y)$, including rates of probability convergence and 'box like' dimensions, and these will be discussed in the following chapters.

It will also be shown that these (and other) probability distributions can also be described by a family of dimensions, known as the Rényi dimensions, which are also related to the multifractal spectrum. In Part II theoretical properties of these dimensions will be discussed and in Part III methods of estimating the Rényi dimensions will be discussed. □

Sets that are irregular can often have different values for different definitions of dimension, e.g., Hausdorff, box, and packing (see Definitions A.2.3, A.3.1 and A.4.2, respectively). The box counting dimension, while it is nice from a conceptual viewpoint, has an unfortunate property, in that the dimension of a set has the same value as the closure of the set. For example, the box dimension of the rational numbers on the real line is one. While the Hausdorff dimension is more difficult to deal with, it has more satisfactory mathematical properties. A summary of definitions and relationships between various dimensions can be found in Appendix A.

The definition of a fractal set originally given by Mandelbrot (1977) was one whose Hausdorff dimension was greater than its topological dimension. Taylor (1986) gives further discussion, and suggests a modification to sets where the Hausdorff and packing dimensions are both equal. If they are not, then the set is so irregular that it may not be able to be described. However, some irregular sets are not fractal according to the above definition. As Stoyan & Stoyan (1994) point out, irregular sets with positive area (e.g., islands) are not fractal, though their boundaries may be fractal (i.e., coastline).

The essential difference between a fractal and a multifractal is that the former refers to a set and the latter refers to a measure. As with fractal sets, a *multifractal measure* may also be extremely irregular with singularities of possibly many different orders. This measure may or may not be supported by a fractal set. One question is: can we partition the support into a multiple number of fractal sets, hence the name multifractal, such that on each individual partition, the order of the singularity of the measure is the same, i.e., the measure is homogeneous, or a unifractal measure on each partition? If this is possible, we would refer to the measure as being multifractal in some sense; the 'sense' being what is precisely meant by 'fractal'. This is essentially what was being done in Example 1.2.1.

We can formalise the above discussion in the context of a probability space $(\mathcal{X}, \mathcal{B}(\mathcal{X}), \mu)$. For our purposes, it will be sufficient to assume that $\mathcal{X} \subset \mathbb{R}^d$, and $\mathcal{B}(\mathcal{X})$ are the Borel subsets of $\mathcal{X}$. In Example 1.2.1, $\mathcal{X} = [0, 1]$, and the support of the Cantor measure μ has dimension (both box and Hausdorff) of $\log_3 2$. The dimension of the support is independent of the manner in which the probability is allocated. We are particularly interested in those situations where the probability measure on $\mathcal{B}(\mathcal{X})$, denoted by μ, is extremely irregular, and the distribution is not differentiable, with singularities of possibly many different orders. Such a measure will be said to be *multifractal*.

Much of the intuition for the development of multifractal theory has originated in the physics literature. Some applications of multifractal theory are discussed in the following sections, in particular, dynamical systems, turbulence, rainfall and earthquakes. In §1.8 we summarise the main features. The chapter concludes by outlining the direction for the remainder of the book.

1.3 Dynamical Systems

Dynamical systems can take the form of an iterative map or a set of differential equations, though the latter can be re-expressed, as follows, in the form of an iterative map. Let $\mathcal{X}$ be a Borel subset of $\mathbb{R}^d$ and consider the transformation

$$T_\xi : \mathcal{X} \longrightarrow \mathcal{X},$$

where ξ is a vector of parameters that modifies the transformation. If $x(t_0) \in \mathbb{R}^d$ is an initial location, then $x(t_n) = T_\xi^n(x(t_0))$, for $n = 1, 2, \cdots$, forms a deterministic sequence of known locations, and is referred to as an *iterative map*. In *discrete time* iterative maps, $t_n = n$. Alternatively, the process may be continuous (in time) and be described by a set of differential equations of the form

$$\frac{dx}{dt} = W_\xi(x(t)),$$

where $t \in \mathbb{R}$, $x(t) \in \mathbb{R}^d$, and ξ is some fixed vector of parameters. Such a process can be approximated by an iterative map by letting $h = t_{n+1} - t_n$ be sufficiently small, then

$$x(t_{n+1}) = T_\xi(x(t_n)) \approx x(t_n) + hW_\xi(x(t_n)) \qquad n \in \mathbb{Z}. \qquad (1.1)$$

Dynamical systems often have one or more of the following characteristics.

1. If $x(t_0)$ is perturbed a very small amount ϵ, the resultant trajectory path

$$x'(t_n) = T_\xi^n(x(t_0) + \epsilon)$$

could diverge and be very different on a point by point basis compared to one starting at $x(t_0)$. Such a system is said to be *chaotic*.

2. Many of these systems have the property that if they are observed for a sufficient length of time, the set that contains the trajectory path of the evolving system will 'look' the same for many different starting values $x(t_0)$. Further, the trajectory path remains within that set, i.e., points within the set are mapped back onto points within the set. This set is referred to as the *attracting set* and is described more precisely by Eckmann & Ruelle (1985). The trajectory path may be periodic, however, this need not necessarily be the case.

3. The transformation T_ξ often has another level of instability, in that as ξ is changed, the characteristics of the system may change through a series of (abrupt) bifurcation points where the shape and other characteristics of the attracting set can change considerably.

Dynamical systems are deterministic, at least theoretically, in that, if we know $x(t_0)$, we can calculate its position at any point in the future. However, from a practical perspective such calculations are generally not possible, due to the finite nature of computer arithmetic. This is interesting, because in chaotic systems (i.e., if $x(t_0)$ is perturbed only a small amount, the resultant trajectory path can be very different) the exact trajectory path could be quite different but the attracting set would be the same.

If $x(t_0)$ is unknown, the process could also be thought of as stochastic. Given that $x(t_0)$ is within the basin of attraction (i.e., will eventually move into the attracting set), then we know that we will 'find' the particle somewhere within that set. Can we describe the probability that the particle is in a set A, i.e., $\mu(A)$? That is, can we observe a process evolving in time, and use this to describe the spatial characteristics of the measure μ? In this situation we need to assume that the measure is invariant under the mapping, i.e., $\mu = T_\xi^{-1}\mu$, and the system is ergodic, so that time averages converge μ-almost all $x(t_0) \in \mathcal{X}$.

There will probably be parts of the attracting set, as seen in the following examples, that are visited frequently and other parts that are visited very infrequently. Effectively, we have a measure μ supported on a set that is possibly very irregular, and the measure itself could also be extremely irregular. One way to describe the size of the attracting set and the spatial characteristics of μ is to calculate various 'generalised dimensions', a method first used and advocated by physicists (see Grassberger, 1983 and Hentschel & Procaccia, 1983).

For the remainder of this section, some examples of dynamical systems are briefly discussed.

1.3.1 The Cantor Map

Consider an infinite sequence (or experimental outcome) of zeros and twos denoted symbolically as $\cdots, \omega_{-2}, \omega_{-1}, \omega_0, \omega_1, \omega_2, \cdots$. Each ω_n is independent with probability p_0 of being zero and $p_2 = 1 - p_0$ of being two. Let

$$x(n) = (0.\omega_n\omega_{n+1}\omega_{n+2}\cdots)_3,$$

where $n \in \mathbb{Z}$ and the right-hand side represents the base 3 fractional expansion (triadic) of $x(n)$. Then $x(n + 1)$ is related to $x(n)$ by a shift operator, i.e.,

$$x(n + 1) = 3x(n) - \lfloor 3x(n) \rfloor,$$

where $\lfloor x \rfloor$ is interpreted as the largest integer not greater than x. This defines a *discrete iterative map* on $[0, 1]$ that has been operating for an infinite amount of time. The Cantor measure of Example 1.2.1 describes the relative proportion of time that the process visits subsets of the unit interval. □

1.3.2 Logistic Map

The logistic map $T_\xi : [0, 1] \to [0, 1]$ is discrete and is defined as $T_\xi(x) = \xi x(1 - x)$, giving a recurrence relation

$$x(n + 1) = \xi x(n)(1 - x(n)), \qquad 0 \le \xi \le 4.$$

If x is a period p point of T_ξ, i.e., $T_\xi^p(x) = x$, and p is the least positive integer with this property, then x is termed *stable* or *unstable* if $|(T_\xi^p)'(x)|$ (i.e., Jacobian) is less than or greater than one, respectively. Stable points attract nearby orbits, unstable points reject them.

There is an interesting sequence of bifurcations occurring with this map as ξ increases to $\xi_\infty \approx 3.57$. When $0 < \xi < \xi_\infty$, T_ξ can have a number of different behaviours: an unstable fixed point at zero, a non-zero stable fixed point, or a stable orbit of period 2^q where $q \in \mathbb{Z}^+$ and $\xi = \xi_q$ with $\xi_q < \xi_{q+1} < \xi_\infty$. When $\xi = \xi_\infty$ the attracting set is of the Cantor type, see Figure 1.4. The attractor is invariant under T_ξ when $\xi = \xi_\infty$. There is no dependence on initial conditions (not chaotic) and the Hausdorff dimension can be estimated as $0.532\cdots$ (see Falconer, 1990, page 173).

In Figure 1.4, it can be seen that the iterates flip from side to side, i.e., will be somewhere in the band between 0.7 and 1.0, then somewhere in the band between 0.3 and 0.7, etc. A series of 200,000 was simulated, and a histogram of the outcomes between 0.4 and 0.6 have been plotted in Figure 1.3 (top histogram). Depending on the 'scale' with which one views the picture, we see a certain number of intervals that appear well populated, and others with no outcomes. More specifically, there are three populated intervals, or clusters of points, each separated from other clusters by a distance of at least 0.03. The second histogram in Figure 1.3 is an enlargement of the first cluster. Enlargements of the first clusters are repeated in the third and last histograms. In each of the enlargements, we notice that the overall structure is the same, both the relative separation between the clusters and the relative number of points in each cluster. Compare this to the scaling characteristics of the Cantor measure when $p_0 = \frac{1}{3}$ in Figure 1.1, where the differences between the probabilities are increasing at each iteration. A similar

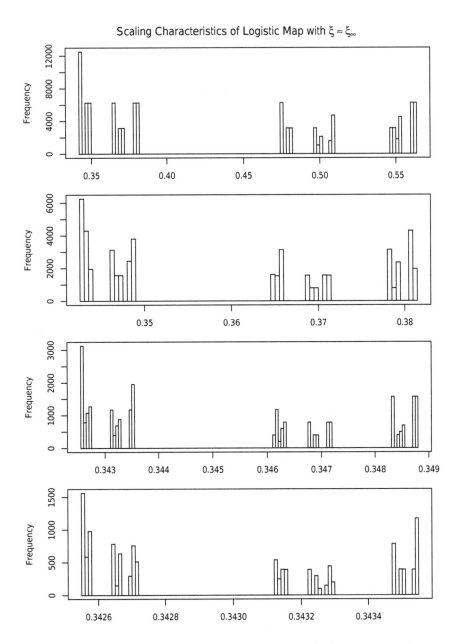

Figure 1.3 *Histograms showing scaling characteristics of the logistic map with* $\xi =$ *3.569945672. The second histogram takes the cluster* < 0.4 *in the first and plots on a finer scale, similarly for the third and fourth histograms. Note that the scaling ratio is approximately* $\frac{1}{6}$.

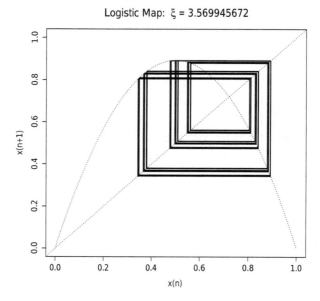

Figure 1.4 *Logistic Map when* $\xi = 3.569945672 \approx \xi_\infty$. *A series of length* 1500 *has been generated with* $x(0) = 0.8$. *Only* $x(t)$ *for* $t = 1001, \cdots, 1500$ *have been plotted, giving the process sufficient time to stabilise to an orbit within the attracting set.*

scaling occurs for the points between 0.7 and 1.0. The histograms in Figure 1.3 suggest that the underlying invariant measure μ is supported on a self-similar like set with scaling parameter that is approximately equal to six.

   May (1976, 1987) suggests using the *logistic map* as a possible model of biological populations, particularly those that die out from generation to generation. This would occur with insects that effectively die out over winter. Such iterative mathematical models were also discussed earlier by Moran (1950) in a more general context. He also discussed when the process would be stable with interpretations in the biological context.   □

1.3.3 Lorenz Attractor (Lorenz, 1963)

The Lorenz time evolution $x(t) = (x_1(t), x_2(t), x_3(t)) \in \mathbb{R}^3$ is defined by the equations

$$W_\xi(t) = \frac{dx}{dt} = \frac{d}{dt} \begin{pmatrix} x_1 \\ x_2 \\ x_3 \end{pmatrix} = \begin{pmatrix} \xi_1(x_2 - x_1) \\ \xi_3 x_1 - x_2 - x_1 x_3 \\ x_1 x_2 - \xi_2 x_3 \end{pmatrix}. \qquad (1.2)$$

Figure 1.5 is plotted using the approximation given by Equation 1.1, and using a first order difference to estimate the derivatives (i.e., Euler's method). With the

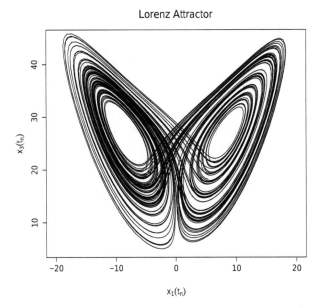

Figure 1.5 *The point* $(1, 1, 1)$ *is transformed using Equation 1.1 with* $h = 0.01$. *The trajectory path from iteration* $1,001$ *to* $11,000$ *is plotted.*

values of $\xi_1 = 10$, $\xi_2 = \frac{8}{3}$ and $\xi_3 = 28$, the trajectories are concentrated onto an attractor of a highly complex form, consisting of two discs of spiraling trajectories. It is chaotic and appears to be fractal, with estimates suggesting a dimension of approximately 2.06. Rotating the plot in three dimensions reveals that the discs are relatively thin, consistent with a dimension of approximately two.

Lorenz (1963) wished to model the thermal convection of a fluid when heated from below, cooling at an upper boundary and then falling, thus circulating in cylindrical rolls. Abarbanel (1995) gives a nice background discussion of the equations and their application in the meteorological context. He notes that the parameter values often used for analysis of the equations are quite different from those which are valid when modelling atmospheric behaviour. Falconer (1990) also gives a brief introduction to the equations from an atmospheric modelling perspective. See also Lorenz (1993), Eckmann & Ruelle (1985, page 622) and Ruelle (1989, page 9) for further discussion.   □

Many other dynamical systems are discussed in the literature. Some examples are the Hénon map (Falconer, 1990; Ruelle, 1989), the bakers' map (Falconer, 1990), Rössler attractor (Ruelle, 1989), Ikeda map (Abarbanel, 1995) and the Kaplan-Yorke map.

The literature on dynamical systems and chaos is vast. There are many descriptive non-mathematical accounts, books containing many quite beautiful pictures of trajectory paths and attracting sets, and others with more detailed technical accounts. Lorenz (1993) gives a nice descriptive overview of chaotic systems. Very good rigorous introductions are provided by Ruelle (1989), Rasband (1990), Abarbanel (1995), and Falconer (1990). Eckmann & Ruelle (1985) is still an important review article, providing a greater depth of detail. Cvitanović (1993), Ott et al. (1994) and Hao (1990) contain collections of reprints of important published papers. The volume by Hao (1990) also contains an extremely extensive bibliography. Ott et al. (1994) also contains a few preliminary chapters providing an introduction to the subject. More mathematical perspectives have often been provided in papers co-authored by David Ruelle. A collection of his papers can be found in Ruelle (1995). Isham (1993) gives a nice introduction from a more statistical perspective. Chatterjee & Yilmaz (1992) review a variety of applications of dynamical systems in different branches of science, and Berliner (1992) discusses the relationship between deterministic chaotic systems and stochastic systems.

1.4 Turbulence

One of the main subject areas that has provided the physical intuition for the development of a theory of multifractals is the desire to describe the nature of energy dissipation in a turbulent fluid flow. Falconer (1990, §18.3) provides a good example of water slowly flowing from a tap where the flow is smooth or *laminar*. As the flow is increased, the flow becomes turbulent or irregular, with 'eddies' at various scales, and varying flow velocities. Cascade models are based on the assumption that kinetic energy is introduced into the system on a large scale (e.g., storms, stirring a bowl of water), but can only be dissipated in the form of heat on very small scales where the effect of viscosity, or friction between particles, becomes important. These models assume that energy is dissipated through a sequence of eddies of decreasing size, until it reaches sufficiently small eddies where the energy is dissipated as heat.

A good historical account of the development of the theory of turbulence is given by Monin & Yaglom (1971), from which much of the following has been drawn. The theory starts with the work of Reynolds in the late 19th century. The Reynolds number is defined as $\mathcal{R} = UL/\nu$, where U and L are characteristic scales of velocity and length in the flow and ν is the kinematic viscosity of the fluid. Therefore, $\mathcal{R}$ is the ratio of typical values of inertial and viscous forces acting within the fluid. The inertial forces produce a transfer of energy from large to small scale components (inhomogeneities), while the viscous forces have the effect of smoothing out the small scale inhomogeneities. Hence flows with a sufficiently small value of $\mathcal{R}$ will be laminar, and sufficiently large may be turbulent.

In the 1920s, Richardson developed a qualitative argument where he assumed that developed turbulence consisted of a hierarchy of 'eddies' (i.e., disturbances or inhomogeneities) of various orders. The eddies arise as a result of a loss of stability of larger eddies, and in turn also lose their stability and generate smaller eddies to which their energy is transferred; hence a cascade type process. Once the scale becomes sufficiently small, then $\mathcal{R}$ is small, hence a laminar flow, with considerable dissipation of the kinetic energy into heat.

Taylor in the 1930s introduced the concepts of homogeneous and isotropic turbulence, determined by the conditions that all the finite dimensional probability distributions of the fluid mechanical quantities at a finite number of space-time points are invariant under any orthogonal transformation. While the assumptions of homogeneity and isotropy are not satisfied in the real situation (e.g., boundary conditions), they provide a useful description of the properties at small scales with a sufficiently high Reynolds number $\mathcal{R}$.

Let $\epsilon(x, t)$ be the rate of energy dissipation per unit volume and unit time at location x and time t. Further, assuming that the turbulence has reached a steady state in time, then the energy dissipation per unit time in a sphere of radius δ centred at x, $S_\delta(x)$, is

$$\mu[S_\delta(x)] = \frac{3}{4\pi\delta^3} \int_{S_\delta(x)} \epsilon(x, t)dx.$$

Kolmogorov (1941) argued that the statistical regime of sufficiently small scale fluctuations of any turbulence with a very high Reynolds number may be taken to be homogeneous and isotropic, and practically steady over a period of time. Thus his assumption that the *average* energy dissipation per unit time is constant over *any* domain. Let this value be denoted by $\bar{\epsilon}$. Note, however, that $\epsilon(x, t)$ is random in nature, and hence the measure of energy dissipation $\mu[S_\delta(x)]$ is also random.

Kolmogorov (1941) further argued that the statistical regime of sufficiently small-scale components of velocity with $\mathcal{R}$ sufficiently large is determined only by $\bar{\epsilon}$ and ν. He argued that the greatest scale that viscosity will still have an effect is $l = (\nu^3/\bar{\epsilon})^{1/4}$. Hence, there is a range many times greater than l but much less than L where the statistical regime is determined by a single parameter $\bar{\epsilon}$. Let $U_{ij}(x)$ be the velocity component (random variable) in the direction $\overrightarrow{x_i x_j}$ at the point x. Kolmogorov then deduced that for arbitrary points x_1 and x_2

$$\mathrm{E}\big[|U_{12}(x_1) - U_{12}(x_2)|^2\big] = c\left(\bar{\epsilon}|x_1 - x_2|\right)^{2/3},$$

where $l \ll |x_1 - x_2| \ll L$, and c is a universal constant.

Subsequently, it was argued that the variation of energy dissipation, $\mu[S_\delta(x)]$, should increase without limit as δ decreases. Kolmogorov (1962) modified the previous '2/3rds law' by assuming that $\log \mu[S_\delta(x)]$ has a normal distribution with a variance that is a function of x and that it increases as δ decreases. This had the effect of treating μ as a random cascade. This argument is used in the literature on rainfall fields discussed below.

As already noted, Monin & Yaglom (1971) provide a detailed historical account of the theory of turbulence. Further discussions can also be found in Mandelbrot (1974), Paladin & Vulpiani (1987), Meneveau & Sreenivasan (1991), Bohr et al. (1998), Frisch (1991), Mandelbrot (1998), and collections of papers contained in Friedlander & Topper (1961) and Hunt et al. (1991).

1.5 Rainfall Fields

A similar cascade argument, to that used for turbulence, has also been applied more recently to rainfall by Schertzer & Lovejoy (1987), Lovejoy & Schertzer (1985, 1990) and Gupta & Waymire (1990, 1993). From Gupta & Waymire (1990):

> These can be viewed generically as 'clusters' of high rainfall intensity rainfall cells embedded within clusters of lower intensity small mesoscale areas, which in turn are embedded within clusters of still lower intensity large mesoscale areas, which are embedded within some synoptic-scale lowest intensity rainfall field.

Gupta & Waymire (1990) give a generalisation to the argument of Kolmogorov (1962) outlined in §1.4. They assume that for a given x, the random measure μ satisfies a more general scaling form given by

$$\mu[S_{r\delta}(x)] \stackrel{d}{=} W(\delta)\mu[S_r(x)] \tag{1.3}$$

for $\delta \leq 1$, and $l \ll r \ll L$, where $\stackrel{d}{=}$ denotes equality of probability distributions, $S_\delta(x)$ is a sphere of radius δ centred at x and $W(\delta)$ is a random function of δ. They show that $W(\delta)$ can be characterised by

$$W(\delta) = \exp\{Z(-\log\delta) + \beta\log\delta\},$$

where $Z(t)$ is a stochastic process with stationary increments, and β is an arbitrary number greater than zero.

1.5.1 Simple Scaling

If the process $Z(t) = 0$ for all t, then $\mu[S_\delta(x)]$ satisfies a simple scaling relation

$$\mu[S_{r\delta}(x)] \stackrel{d}{=} \delta^\beta \mu[S_r(x)]. \tag{1.4}$$

Let $r = 1$, and take expectations on both sides (i.e., of μ) as follows to give

$$\log \mathrm{E}[\mu^q[S_\delta(x)]] = q\beta\log\delta + \log \mathrm{E}[\mu^q[S_1(x)]]$$

where $q \in \mathbb{R}$. Assuming that $\mathrm{E}[\mu^q[S_1(x)]]$ is finite, then for sufficiently small δ,

$$\frac{1}{q-1}\frac{\log \mathrm{E}\left[\mu^{q-1}[S_\delta(x)]\right]}{\log\delta} \approx \beta. \tag{1.5}$$

1.5.2 Example - Brownian Multiplier

The model considered by Kolmogorov (1962) and Oboukhov (1962) was a special case of that in Equation 1.3. They assumed that $Z(t) = \sigma B(t)$, where $\sigma > 0$ and $B(t)$ is Brownian motion. Now consider the following heuristic argument. Assume that $W(\delta)$ is independent of $\mu[S_r(x)]$. Let $r = 1$, and take expectations on both sides (i.e., of μ and W) to give

$$\log \mathrm{E}[\mu^q[S_\delta(x)]] = \log \mathrm{E}[\exp(q\sigma B(-\log\delta) + q\beta\log\delta)] + \log \mathrm{E}[\mu^q[S_1(x)]]$$

where $q \in \mathbb{R}$. Assuming that $\mathrm{E}[\mu^q[S_1(x)]]$ is finite, then for sufficiently small δ,

$$\frac{1}{q-1} \frac{\log \mathrm{E}\left[\mu^{q-1}[S_\delta(x)]\right]}{\log \delta} \approx -\frac{(q-1)\sigma^2}{2} + \beta. \tag{1.6}$$

□

The left-hand sides of Equations 1.5 and 1.6 have the form of a Rényi dimension of order q, to be defined in Chapter 2. It can be seen from Equation 1.5 that, in the case of simple scaling, all Rényi dimensions are the same; whereas from Equation 1.6, in the case where the scaling contains a stochastic component, the Rényi dimensions are all different. Similar behaviour occurs with the Cantor measure of Example 1.2.1. It will be shown in Example 2.2.1 that if the probability is allocated unevenly as in Figure 1.1, then the Rényi dimensions will be different; whereas if the probability is allocated evenly (i.e., $\frac{1}{2}$ to each subinterval), then the Rényi dimensions will all be the same.

There is a subtle distinction between the random measures of this section and the Cantor measure as constructed in Example 1.2.1. The Cantor measure is constructed in a deterministic iterative manner, whereas a random measure is more complicated and is constructed in an iterative stochastic manner. The Cantor measure is a special case of the family of multinomial measures, which will be discussed more fully in Chapter 3. The random measures discussed above in the context of rainfall and turbulence are examples of random cascades, and will be discussed more fully in Chapter 6.

The simple scaling relation given by Equation 1.4 is quite similar to that of a self-similar stochastic process (Samorodnitsky & Taqqu, 1994). A stochastic process $X(t)$ is said to be *self-similar* if its finite dimensional distributions satisfy the scaling relation

$$X(\delta t) \overset{d}{=} \delta^H X(t) \tag{1.7}$$

for all $\delta > 0$, $t \in \mathbb{R}$ and $0 < H < 1$ (see §10.6.2 for further details). An example of such a process is the increments of fractional Brownian motion (Mandelbrot & Van Ness, 1968). When $H > \frac{1}{2}$ the process displays long range dependence (Beran, 1994), when $H = \frac{1}{2}$ the increments of fractional Brownian motion are simply white noise, and when $H < \frac{1}{2}$ there is short range dependence with negative autocorrelations. More recently, work has been done on stochastic processes satisfying a more general analogue of the scaling relationship in Equation 1.3, i.e.,

$$X(\delta t) \overset{d}{=} W(\delta)X(t).$$

These processes are peripheral to the main direction of material contained in this book and will be discussed only briefly in §10.6.

1.6 Earthquake Modelling

A number of the world's larger cities are located in seismically active zones. Loss of life caused by earthquakes during the 20th century was immense. As recently as 1976, approximately 240,000 people died in the Tangshan, China, earthquake (28 July 1976). Other large events with a corresponding large loss of life also occurred in San Francisco (18 April 1906; 3,000 died), Tokyo (1 September 1923; 140,000 died), Mexico City (19 September 1985; 8,000 died), and Izmit, Turkey (17 August 1999; 17,000 died).

Earthquake prediction and forecasting has had periods of very active research in the last 100 years. In the 1970s, there was a great deal of optimism in the scientific community that *individual earthquake events* could be predicted. This coincided with considerable developments in the understanding of the earth's structure (plate tectonics) and also in better catalogues of located earthquake events. This optimism quickly diminished though, and some scientists now believe that the problem is so complex that individual earthquake events cannot be predicted (see Aki, 1989; Kagan, 1997; Geller, 1997; Geller et al., 1997; Wyss et al., 1997; and Kagan, 1999). However, many still hope that useful forecasts of relative probabilities can be made. These may consist of contour maps, rather like those of the weather maps, comparing the relative probabilities of an event greater than a given magnitude, in different areas or regions. Vere-Jones (1995) gives a review of earthquake forecasting and Vere-Jones (2000) gives a brief introduction to seismology from a statistical perspective. More detailed general accounts of the subject are provided by Lay & Wallace (1995) and Scholz (1990).

Mathematical models that describe the fracturing process are relatively primitive compared to models that describe the evolution and behaviour of weather systems. Some models postulate that there are 'elementary dislocations' occurring all of the time. Periodically, the occurrence of a number of these dislocations will cause a cascade of further elementary dislocations. If the cascade is sufficiently large, an earthquake will be detected by a sufficiently sensitive seismic network.

There are a number of powerlaw relationships describing seismicity that empirical evidence tends to support. These include the magnitude distribution of events (Gutenberg-Richter law) and the decay over time in the number of events after a large mainshock (Omori's law). The intuitive motivation for estimating the fractal dimension of spatial point patterns generated by earthquakes is that the pattern may be self-similar in some sense. That is, clusters may be repeated within clusters on a finer and finer level (see, for example, Figure 6.1). Though some clusters may be more active than others, in the same way that the Cantor measure is not necessarily uniform over its support. It was also thought that major fractures occur along major faults, the most dramatic being the tectonic plate boundaries. Within major fault systems there are smaller faults that branch off, and from these smaller fault networks; again with the possibility of generating some sort of self-similar hierarchy of networks.

Dimension estimates in the earthquake context are primarily descriptive in nature. If the earthquake process really did display fractal like characteristics, then it would be desirable for one's models of the fracture process and those for forecasting of event probabilities also to display similar fractal characteristics.

However, there is an inherent contradiction in calculating dimensions of point patterns. A finite set of point locations theoretically has dimension zero. Hence, what characteristics are the dimension estimates describing? This question depends on the underlying model that one has in mind, and will be discussed further in Chapter 11.

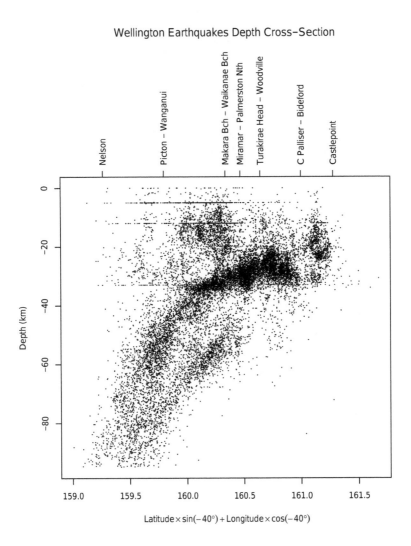

Figure 1.6 *Depth cross-section of Wellington earthquake locations between 1985 and 1994, with magnitude $\geq$ 2 and depth $<$ 95 km. The plot contains 15,410 events. The picture shows the Pacific Plate subducting the Australian Plate. The angle of view is (approximately) from the southwest to the northeast.*

1.6.1 Wellington Earthquake Catalogue

The Wellington Earthquake Catalogue contains events from an area of central New Zealand. For our analyses in Chapter 11, events have been selected with magnitude ≥ 2 between 173.6°E and 176°E, 42.0°S and 40.4°S, and occurring between 1 January 1978 and 31 December 1995. The catalogue is maintained by the Institute of Geological and Nuclear Sciences, Wellington (see Maunder, 1994).

The surface of the earth consists of large tectonic plates, for example the North American, Eurasian, Pacific and Australian Plates (see Sphilhaus, 1991). Most earthquake activity in the world is located in the vicinity of these plate boundaries, and is caused by the movement of one plate relative to the other. New Zealand is located on the boundary of the Australian and Pacific tectonic plates. In the Wellington Region, the Pacific Plate *subducts* the Australian Plate, that is, the Pacific Plate is drawn beneath the Australian Plate. The two lines of events in Figure 1.6 roughly mark the location of the friction boundary of the subducting

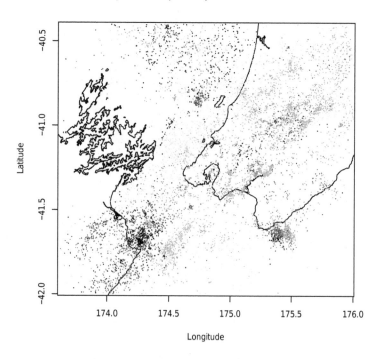

Figure 1.7 *Wellington earthquake epicentres between 1985 and 1994, with magnitude ≥ 2 and depth < 40 km. The deepest events are in the lightest shade of gray and the most shallow events are the darkest. The plot contains* 10,801 *events.*

Wellington Earthquake Epicentres: Deep Events

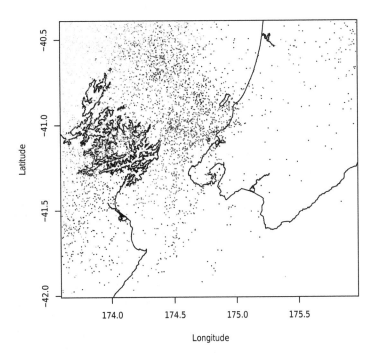

Figure 1.8 *Wellington earthquake epicentres between 1985 and 1994, with magnitude ≥ 2 and depth $\geq 40\,km$. The deepest events are in the lightest shade of gray and the most shallow events are the darkest. The plot contains 4,952 events.*

Pacific Plate. It can also be seen that most of the events with a depth $\geq 40\,km$ are associated with the subduction process, whereas those more shallow events appear to have a more widespread distribution. We use $40\,km$ as a boundary between shallow and deep events for the Wellington Catalogue. The lines of events occurring at 5, 12 and 33 km mark shallow events with a poorly determined depth.

Figures 1.7 and 1.8 are epicentral plots of shallow and deep events, respectively. The subduction process is also evident in Figure 1.8, where the deeper events tend to occur to the northwest. Notice also that the shallow events appear to be more clustered, and spread more widely over the region. □

There are many problems in estimating fractal dimensions using 'real' data. For example, the earthquake locations contain location errors which may not even be homogeneous over the analysed region. There are also boundary effects caused by the inability of the seismic network to accurately detect events that are too distant. These problems are discussed more fully in the analyses of Chapter 11.

1.7 Other Applications

Many of the applications of multifractal measures have been to describe physical processes like turbulence, dynamical systems, rainfall and earthquakes. Paladin & Vulpiani (1987) give a review of the application of multifractals to a number of fields in physics, including turbulence. Scholz & Mandelbrot (1989) contain a collection of papers on the application of fractal ideas in geophysics.

Some have also postulated that global climate is determined by a chaotic dynamical system of relatively low dimensionality (see Nicolis & Nicolis, 1984). An observed time series, like daily maximum temperatures in Wellington, NZ, could be thought of as a projection of this dynamical system to an observation space of lower dimension. Using such observed data, is it possible to determine the 'dimension' of the global system? This problem will be briefly discussed in §10.4.4. If the process is not deterministic, but stochastic, then we would expect it to have an infinite number of degrees of freedom (see §10.5). Using local temperature data, Wang (1995) attempted to determine whether there was evidence that these data were generated by a system of relatively low dimensionality. He concluded that his estimates of low dimensionality could be explained by statistical biases.

Parallel developments to the theory of multifractal measures have occurred with self-similar stochastic processes (Equation 1.7). The increments of these processes can be distributed as a stable law, or may have long range dependence. Various associated *auxiliary processes* (level crossings, etc.) have fractal like characteristics and will be discussed only briefly in §10.6. These processes have been used to model financial data (Mandelbrot, 1997) and internet traffic (Willinger et al., 1995; Resnick, 1997; Willinger & Paxson, 1998; and Park & Willinger, 2000).

Self-similar stochastic processes are somewhat peripheral to the material we discuss in this book because the primary focus of interest with these models is in their long range dependence and heavy tail characteristics. The fractal characteristics of the associated auxiliary processes appear to be of only secondary importance. Further, self-similar stochastic processes only satisfy a monoscaling law. More recently, the monoscaling aspect of self-similar stochastic processes has been extended to processes that are multiscaling, which have been referred to as 'multifractal' stochastic processes. These will be briefly discussed in §10.6.5.

1.8 Concept of Multifractals

In this chapter, examples of dynamical systems, turbulence, rainfall processes and earthquake events have been briefly discussed. Consider these examples in the context of a measure space $(\mathcal{X}, \mathcal{B}(\mathcal{X}), \mu)$. In the case of the Cantor measure in Example 1.2.1, $\mathcal{X} = [0, 1]$, and the support of the measure μ was the Cantor set which has a 'fractal' dimension of $\log_3 2$ (both box and Hausdorff). In the case of the dynamical systems, $\mathcal{X} \subseteq \mathbb{R}^d$ is the phase space. The measure $\mu(A)$ can be thought of as the probability of the set $A \in \mathcal{B}(\mathcal{X})$ containing the trajectory at any given instant. In the discussion about turbulence, μ was a measure of energy dissipation, and was assumed to be random in nature.

In each of these examples, the support of the measure μ could be extremely irregular and have zero Lebesgue measure. The measure itself could also be extremely variable and have singularities of many different orders. As such, these measures do not have an associated density. How does one characterise such measures? Many of these measures have multifractal like properties. An intuitive introduction to multifractals is given in this section. Some of the concepts are rather loosely defined, and the material will be covered more rigorously in later chapters.

Consider a lattice covering of $\mathcal{X}$ by d-dimensional boxes of width δ_n, where $B_{\delta_n}(x)$ is the box that contains the point x. The sequence $\delta_n \to 0$ as $n \to \infty$. Then consider another mapping

$$U_n : \mathcal{X} \longrightarrow \mathbb{R},$$

where $U_n(x) = -\log \mu[B_{\delta_n}(x)]$ if $\mu[B_{\delta_n}(x)] > 0$. Let $Y_n(x)$ be a rescaled version of $U_n(x)$, i.e.,

$$Y_n(x) = \frac{U_n(x)}{-\log \delta_n},$$

The variable Y_n describes the *local* or individual box behaviour of the measure μ.

In most situations, the limiting probability distribution of Y_n is trivial, that is, there exists a number y_0 such that

$$\lim_{n \to \infty} \Pr\{Y_n = y\} = 0 \quad \text{if } y \neq y_0.$$

Another way to characterise Y_n is to describe the rate at which the probability tends to zero, i.e., the number of boxes where Y_n has some given rate (as in Example 1.2.1). This is referred to as the *multifractal spectrum*, $\widetilde{f}(y)$, which can be expressed as

$$\widetilde{f}(y) = \lim_{\epsilon \to 0} \lim_{n \to \infty} \frac{\log\{\text{box count at } n\text{th stage with +ve } \mu \text{ and } |Y_n - y| < \epsilon\}}{-\log \delta_n}.$$

While $\widetilde{f}(y)$ is similar to a box counting dimension, it is not necessarily the same, as the set that is being covered by boxes of decreasing widths is also changing its nature as $n \to \infty$. Let N_n be the number of boxes at the nth stage with positive μ measure. Then

$$\widetilde{f}(y) = \lim_{n \to \infty} \frac{\log N_n}{-\log \delta_n} + \lim_{\epsilon \to 0} \lim_{n \to \infty} \frac{\log \Pr\{|Y_n - y| < \epsilon\}}{-\log \delta_n}. \qquad (1.8)$$

The first term is the box counting dimension of the support of μ. Given that $y \neq y_0$, the second term is the powerlaw rate that the probability function of Y_n approaches zero. It is this second term that the theory of large deviations focuses on describing, and in that context is often referred to as the *entropy* function.

Now consider a rescaled cumulant generating function of $U_n(x)$, denoted by $\widetilde{\theta}(q)$, as

$$\widetilde{\theta}(q) = \lim_{n\to\infty} \frac{\log \mathrm{E}[\exp(-(q-1)U_n(X))]}{\log \delta_n} = \lim_{n\to\infty} \frac{\log \mathrm{E}\left[\mu^{q-1}[B_{\delta_n}(X)]\right]}{\log \delta_n}.$$

Here the expectation acts on the random variable X with probability distribution given by μ. In the case of the processes discussed in §1.4 and §1.5, μ is the random variable, but the form of the function is essentially the same. It should be noted that $\widetilde{\theta}(q)$ will appear in two different contexts. Firstly, the Rényi dimensions are $\widetilde{D}_q = \widetilde{\theta}(q)/(q-1)$. Recall that these occurred in Equations 1.5 and 1.6, simple scaling and more complex scaling, respectively. In the case of simple scaling, all Rényi dimensions were the same, whereas they were different in the case of complex scaling. Also recall that the dimension of the support of the Cantor measure in Example 1.2.1 is invariant to the way that the probability is allocated. It is these Rényi dimensions that change according to the manner in which the probability is allocated within the Cantor set. The second context is that the rescaled cumulant generating function is a type of global average and occurs in the theory of large deviations, which will be used in Part II of the book.

The *global averaging* or Rényi dimensions, given by $\widetilde{\theta}(q)$ or $\widetilde{D}_q$, respectively, are often related to the multifractal spectrum, $\widetilde{f}(y)$, by a Legendre transform. We are interested when such a relationship holds. In these situations, we will think of μ as a multifractal measure in a *weak sense*; formal definitions will be given in Chapter 2. The same relationship often holds between the rescaled cumulant generating function and what is called the *entropy* function (i.e., the last term on the right-hand side of Equation 1.8) in the theory of large deviations. We will therefore use that theory to determine necessary conditions for the measure μ to be a multifractal measure in a weak sense.

For some measures it can be shown that $\widetilde{f}(y)$ has a considerably stronger interpretation, that is, $\widetilde{f}(y) = \dim_H \widetilde{F}(y)$, where

$$\widetilde{F}(y) = \left\{x : \lim_{n\to\infty} Y_n(x) = y\right\} = \left\{x : \lim_{n\to\infty} \frac{\log \mu[B_{\delta_n}(x)]}{\log \delta_n} = y\right\},$$

and $\dim_H$ is the Hausdorff dimension. Hence in this case, the partition alluded to in §1.2 is $\widetilde{F}(y)$, and the 'fractal' dimension is the Hausdorff dimension. From Mandelbrot (1989), who uses the notation $f(\alpha)$ in place of $\widetilde{f}(y)$:

> A multifractal measure can be represented as the union of a continuous infinity of addends. Each addend is an infinitesimal 'unifractal measure'. It is characterised by a single value of α, and is supported by a fractal set having the fractal dimension $f(\alpha)$. The sets corresponding to the different α's are intertwined.

Describing the measure μ by lattice coverings is not the only possible method. One may also describe local behaviour by considering

$$\frac{\log \mu[S_\delta(x)]}{\log \delta}$$

where $S_\delta(x)$ is a sphere of radius δ centred at the point x. An analogous definition to $\widetilde{\theta}(q)$ for global averaging can also be made. We will refer to this framework as a *point centred* construction.

The case described above is where μ is a probability measure. Another possibility is that μ is a random measure constructed by a cascade process. This case arises in the context of turbulence, which was discussed in §1.4 .

We have briefly outlined above two methods of characterising multifractal behaviour: one with a *lattice* based construction and the other with a *point centred* construction. Both relate *local* behaviour to *global* averaging. Further, 'fractal' dimension can be interpreted in a *weak* or *strong* sense. In the literature, these are often referred to as *coarse* and *fine* grained, respectively (see Falconer, 1997). Definitions of multifractal measures, and methods of construction will be given in Chapter 2.

1.9 Overview of Book

The book is split into three parts.

Part I - Introduction and Preliminaries

In this chapter we have described various characteristics of a multifractal measure, denoted by μ. The measure could be constructed in a deterministic manner as for the Cantor set in Example 1.2.1, or may be random as in §1.4 and §1.5. It could be supported by a fractal set whose dimension will be invariant to the manner in which the measure is allocated within the set. We can describe the distribution of the measure by investigating its local or global behaviour, and in some situations these will be related by a Legendre transform. The same relationships hold in the theory of large deviations (i.e., between $\widetilde{\theta}(q)$ and the entropy term in Equation 1.8). The Rényi dimensions are based on the global (averaging) behaviour. Estimates of 'fractal' dimensions in empirical studies are, in general, estimates of these Rényi dimensions. From such estimates it is possible, at least theoretically, to estimate the multifractal spectrum $\widetilde{f}(y)$.

In this chapter, various technical terms have been used in a sometimes rather loose manner. In Chapter 2 definitions of a multifractal measure in both a weak and strong sense, and also using lattice based and point centred spherical constructions, will be defined. In the case of the multinomial measures, which the Cantor measure is a special case, the Legendre transform relationship between the rescaled cumulant generating function $\widetilde{\theta}(q)$ and the multifractal spectrum $\widetilde{f}(y)$ can be demonstrated relatively easily using Lagrange multipliers. This example will be discussed in Chapter 3.

Part II - Multifractal Formalism Using Large Deviations

In Part II, we use the Gärtner-Ellis theorem of large deviations as a foundation
to provide a parallel development of lattice based and point centred constructions
in Chapters 4 and 5, respectively. The Gärtner-Ellis theorem of large deviations
and other related results are reviewed in Appendix B. Note that we use tildes on
$\theta(q)$ and $f(y)$ in the lattice based constructions and no tildes in the point centred
construction. We want to determine fairly general conditions under which the
global and local behaviours of a measure are related via a Legendre transform. We
are also interested in determining under what conditions the multifractal spectrum
can be interpreted as a Hausdorff dimension.

Given the lattice based and point centred constructions, under what conditions
do the global and local behaviours of the two constructions coincide? This question will be discussed at the end of Chapter 5. Part II concludes with Chapter 6,
which reviews other cascade constructions, some of which cannot be satisfactorily
described using the framework of large deviations.

Part III - Estimation of the Rényi Dimensions

In Part III, the emphasis is on the estimation of the Rényi dimensions. In the
preceding discussion, we have referred to both lattice based and point centred
multifractal constructions. In the context of estimation, both are different. If estimating the Rényi dimensions in a lattice based situation, one would cover by
a lattice system of boxes, counts taken, and averaged accordingly. In the point
centred case, one analyses interpoint distances. In fact $E\left[\mu^{q-1}[S_\delta(X)]\right]$, which
is the expectation term that appears in the rescaled cumulant generating function
in the point centred situation, is simply the probability distribution function of an
interpoint distance of order q (say Y) when $q = 2, 3, \cdots$; i.e., $\Pr\{Y \leq y\} =$
$F_Y(y) = E\left[\mu^{q-1}[S_y(X)]\right]$. An interpoint distance of order q (i.e., Y) will be
defined in Chapter 2. Hence the Rényi dimensions are essentially the powerlaw
exponent of the probability function $F_Y(y)$ assuming such an exponent exists,
i.e., $F_Y(y) \sim y^{\theta(q)}$. At this point, the problem looks relatively easy, one simply
draws a sample of many interpoint distances of order q, estimates $\theta(q)$, doing
it for a number of values of q. One can then partially reconstruct the multifractal spectrum. Unfortunately the problem is not quite as easy, mainly because of
various forms of bias in the estimates of $\theta(q)$.

In Chapter 7, the correlation integral is defined and related back to the definition
of Rényi dimensions in Chapter 2. Intrinsic features of the correlation integral
can cause bias in the estimation of the correlation exponents, and are discussed
in Chapter 7. These are particularly evident when the measure is supported on
certain self-similar sets. This causes the function $F_Y(y)$ to have an oscillatory
like behaviour which is periodic on a logarithmic scale. This means that $F_Y(y)$
only has powerlaw behaviour in an average sense.

We have generally used a modified Hill estimator for estimating the Rényi dimensions. This is described in Chapter 8, along with some other methods of estimation. In Chapter 9, we describe various extrinsic sources of bias, that is, not inherent to the correlation integral itself, but possibly due to sampling strategies, and other data handling deficiencies. These errors are analogous to non-sampling error in sample surveys. The three main problems here are noise or error in the data, rounding of data, and the boundary effect. The boundary effect often occurs in estimation methods that are based on interpoint distances. Both the rounding and the noise in the data have the effect of blurring out the fine scale information. Since dimensions are a limiting concept as interpoint distances become very small, these forms of bias can be quite serious.

In Chapter 10, the Rényi dimensions are estimated using data that have been simulated from various statistical and mathematical models. In some of these models, the dimensions can be calculated analytically, and in others, estimates have been made by many researchers, and there is some consensus on what the actual values are. Even in these analyses, both the intrinsic and extrinsic forms of bias are evident. These forms of bias also need to be disentangled from those of the estimator itself. Using models with at least partially understood properties helps in understanding general estimation problems.

Part III concludes with Chapter 11, where Rényi dimensions are estimated using earthquake hypocentre locations of events occurring in New Zealand and Japan. These data are interesting, not only from the perspective of earthquake forecasting, but also because they contain many of the forms of bias discussed in Part III of the book.

CHAPTER 2

The Multifractal Formalism

2.1 Introduction

In this chapter we give definitions of the Rényi dimensions and multifractal spectrum in both lattice based and point centred settings. Definitions are also given of the so called multifractal formalism. Related results from the literature are also reviewed.

Let $\mathcal{X}$ be a Borel subset of $\mathbb{R}^d$ and $\mathcal{B}(\mathcal{X})$ be the Borel sets of $\mathcal{X}$. We are interested in the probability space $(\mathcal{X}, \mathcal{B}(\mathcal{X}), \mu)$, where the non-atomic probability measure μ may be concentrated on a subspace of $\mathcal{X}$ of lower dimension than d. For example, in the case of the Lorenz attractor (§1.3.3), $d = 3$, however, it appears that the attracting set is mostly concentrated on two discs (see Figure 1.5), which empirical studies indicate have dimension slightly greater than 2. The measure $\mu(A)$ gives the probability of finding the trajectory in the set A at any given time.

Note that if we wish to describe the spatial characteristics of a trajectory path based on observations of the process over time, then assumptions of invariance and ergodicity are required. The notion of invariance means that the measure μ is unchanged under the transformation given by Equation 1.1, i.e., $\mu = \mu T_\xi^{-1}$. Ergodicity means that averages over time (i.e., averages of repeated operations of T_ξ) are the same as the corresponding spatial averages for μ-almost all x. Further detailed discussion of these ideas can be found in the texts by Walters (1982) and Billingsley (1965).

We consider two multifractal constructions.

1. The case where $\mathcal{X}$ is covered by a succession of lattices of d-dimensional boxes of diminishing width δ_n as $n \to \infty$. We will refer to this as the *lattice* case.

2. The case where $\mu[S_\delta(x)]$ is analysed for all x such that $\mu[S_\delta(x)] > 0$, where $S_\delta(x)$ is a closed sphere of radius δ centred at x. We will refer to this as the *point centred* case.

Functions that relate to the lattice case, and may be confused with the point centred case, will be over struck with a tilde.

In both cases, we want to describe *global* and *local* behaviour. The underpinning concepts of global behaviour are based on the work by the Hungarian mathematician Alfréd Rényi on information theory. A brief review of this work is given in §2.2, which forms the basis of the Rényi dimensions in §2.3 and §2.4. Local behaviour is described by the multifractal spectrum, to be defined in §2.5.

The multifractal formalism involves a relationship between global (Rényi dimensions) and local behaviours (multifractal spectrum) in the form of a Legendre transform. We will refer to this as the *weak* case. The *strong* case is where the multifractal spectrum can be interpreted as a family of Hausdorff dimensions. Definitions of the multifractal formalisms in both *weak* and *strong* senses will be given in §2.5.

Our definitions are not the only way to describe the multifractal formalism. There are various definitions, each with their inherent difficulties and weaknesses. Sections 2.6 and 2.7 review these problems and other results that relate to lattice based and point centred constructions respectively.

Note the difference between the random measures alluded to in the discussion about turbulence in §1.4 and the probability measures being discussed in this chapter. In this chapter and Chapters 4 and 5, μ will always be a probability measure. The discussion about random measures started in §1.4 will be taken up again in Chapter 6.

2.2 Historical Development of Generalised Rényi Dimensions

The Rényi dimensions originate from information theory. This theory arises in connection with the transmission of information, in particular, the length of a binary representation of that information. Say a set E has n elements. If $n = 2^N$, where $N \in \mathbb{Z}^+$, each element can be labelled by a binary number having N digits. As such, Hartley defined $\log_2 n$ as the necessary information to characterise E.

Now assume that $E = E_1 + E_2 + \cdots + E_b$, where $E_1, \cdots, E_b$ are pairwise disjoint finite sets. An experiment is performed, which consists of independently and randomly allocating the n elements to the b subsets (E_k, $k = 1, \cdots, b$) according to the probabilities p_k. The amount of information generated by such an experiment about the probability distribution $P = (p_1, \cdots, p_b)$ is

$$H_1(P) = -\sum_{k=1}^{b} p_k \log_2 p_k.$$

This is known as *Shannon's formula*.

Rényi (1965) showed that Shannon's formula can also be derived as follows. Assume that there is a particular element of interest, however, we do not know which of the n elements it is. A sequence of the above experiments are performed, where the n elements are independently and randomly allocated to the b subsets (E_k, $k = 1, \cdots, b$) according to the probabilities p_k. We are only told after each experiment which subset each element is allocated to and which subset contains the unknown element of interest. The first experiment produces a partition Δ_1, the 2nd experiment produces a partition Δ_2, etc. Let $\Delta^{(m)}$ denote the cross product of the m partitions generated by the first m independent experiments. Each element can be thought of as taking a path of length m through $\Delta^{(m)}$. The unknown element will be determined uniquely when its path is unique with respect to all other elements. Let $P^\star_{nm}$ be the probability that this unknown element can be uniquely identified after m experiments, i.e., its path is unique. Define

$$e_1(n, \epsilon) = \min \{m : P^\star_{nm} \geq 1 - \epsilon\},$$

where $0 < \epsilon < 1$, then Rényi showed that

$$\lim_{n \to \infty} \frac{\log_2 n}{e_1(n, \epsilon)} = H_1(P).$$

A heuristic interpretation is as follows. The total amount of information required to characterise E is $\log_2 n$. Each experiment gives $H_1(P)$ information, therefore, if n is sufficiently large, approximately $(\log_2 n)/H_1(P)$ experiments are required.

An axiomatic approach can also be taken to determine the form of the function $H_1(P)$, where it can be shown that it is the only function that has the required properties. One such property is the additive property. Consider two different probability distributions with which to partition E,

$$P = (p_1, \cdots, p_b) \quad \text{and} \quad Q = (q_1, \cdots, q_a).$$

Let $P \star Q$ be the distribution of the terms $p_j q_i$, $i = 1, \cdots, a; j = 1, \cdots, b$, then H_1 satisfies an additive property, i.e.,

$$H_1(P \star Q) = H_1(P) + H_1(Q). \tag{2.1}$$

Using the above context, Rényi extended the notion of information to higher orders (Rényi, 1965). Let $P^{(q)}_{nm}$ denote the probability that each class of $\Delta^{(m)}$ contains less than q elements, i.e., each possible path of length m contains less than q elements taking the same route. Let

$$e_q(n, \epsilon) = \min \left\{m : P^{(q)}_{nm} \geq 1 - \epsilon\right\} \qquad q = 2, 3, \cdots,$$

then Rényi showed that

$$\lim_{n \to \infty} \frac{\log_2 n}{e_q(n, \epsilon)} = \left(1 - \frac{1}{q}\right) H_q(P) \qquad q = 2, 3, \cdots,$$

where

$$H_q(P) = \frac{-1}{q-1} \log_2 \sum_{k=1}^{b} p_k^q.$$

Further, $\lim_{q \to 1} H_q(P) = H_1(P)$ and $H_q(P)$ satisfies the additive property of Equation 2.1. As with $H_1(P)$, the functional form of $H_q(P)$ can be argued from a pragmatic perspective or from an axiomatic approach (see Rényi, 1965, 1970). See Rényi (1970, page 581) for further discussion of the case where $q < 0$.

Rényi (1959) introduced the idea of dimension as follows. Consider a random variable X taking countably many values x_k with probability $p_k = \Pr\{X = x_k\}$.

Then we could define the information contained in the value of X as

$$H_1(P) = -\sum_{k=1}^{\infty} p_k \log_2 p_k$$

and

$$H_q(P) = \frac{-1}{q-1} \log_2 \sum_{k=1}^{\infty} p_k^q \qquad q \neq 1.$$

However, when the distribution of X is continuous, the amount of information would be infinite, assuming the value of X can be determined exactly.

Rényi (1959) then described X in terms of a discrete random variable $X_n = \lfloor nX \rfloor / n$, where $\lfloor x \rfloor$ denotes the integer part of x, and investigated the behaviour as $n \to \infty$. Let $P_n = (p_1, p_2, \cdots)$ be the probability distribution of X_n. Then he defined the *dimension* of the distribution of X as

$$d_q = \lim_{n \to \infty} \frac{H_q(P_n)}{\log_2 n}.$$

This is telling us how fast the information of X is tending to infinity. See Rényi (1970, page 588) for further discussion of this dimension.

2.2.1 Example

Let X be a random variable that is sampled from a distribution given by the Cantor measure as in Example 1.2.1. Consider a similar situation as in Example 1.3.1, though where we have an infinite one sided sequence $\omega_1, \omega_2, \cdots$ of zeros and twos with probabilities p_0 and p_2. Consider the random variable X, written with a base 3 fractional representation (triadic) as $X = (0.\omega_1\omega_2\omega_3 \cdots)_3$.

Then let

$$
\begin{aligned}
X_1 &= (0.\omega_1)_3, \\
X_2 &= (0.\omega_1\omega_2)_3, \quad \text{and} \\
&\vdots \\
X_n &= (0.\omega_1\omega_2 \cdots \omega_n)_3 = \frac{\lfloor 3^n X \rfloor}{3^n}.
\end{aligned}
$$

At each step, one more digit in the triadic expansion of X is revealed. At what rate are we accumulating information about the Cantor measure μ as n increases?

Let P_n be the probability distribution of the discrete random variable X_n, which can take 2^n possible values, say x_k where $k = 1, \cdots, 2^n$. In the case where $q \neq 1$,

$$
\begin{aligned}
H_q(P_n) &= \frac{-1}{q-1} \log_2 \sum_{k=1}^{2^n} (\Pr\{X_n = x_k\})^q \\
&= \frac{-1}{q-1} \log_2 (p_0^q + p_2^q)^n \\
&= \frac{-n}{q-1} \log_2 (p_0^q + p_2^q).
\end{aligned}
$$

Hence the Rényi dimensions for $q \neq 1$ are

$$
d_q = \lim_{n \to \infty} \frac{H_q(P_n)}{\log_2(3^n)} = \frac{-1}{q-1} \log_3 (p_0^q + p_2^q).
$$

Similarly, when $q = 1$,

$$
d_1 = \lim_{n \to \infty} \frac{H_1(P_n)}{\log_2(3^n)} = -(p_0 \log_3 p_0 + p_2 \log_3 p_2).
$$

In Example 1.2.1 we described the size of the support of the Cantor measure by its dimension, though this was invariant to the way that the measure was allocated within the support. Using the Rényi dimensions, we can also characterise the way that the measure is allocated. In the following sections, we define the Rényi dimensions in a more general context. □

2.3 Generalised Rényi Lattice Dimensions

Consider a lattice covering of the support of μ by d-dimensional boxes of width δ_n that are half open to the right, usually with a node anchored at the origin. The kth box is denoted by $B_{\delta_n}(k)$, where $k \in K_n$ and $K_n = \{k : \mu[B_{\delta_n}(k)] > 0\}$. We evaluate successive lattice coverings for some sequence $\{\delta_n\}$, where $\delta_n \to 0$ as $n \to \infty$.

In the context of information theory, as developed by Rényi, we could think of the situation as follows. Let $\mathcal{F}_n$ be the σ-field generated by all boxes $B_{\delta_n}(k)$, where $k \in K_n$. We are then interested in the probability space $(\mathcal{X}, \mathcal{F}_n, P_n)$, where $P_n(A) = \mu(A)$ for $A \in \mathcal{F}_n$. Summing over all possible outcomes is equivalent to summing over all lattice boxes. Hence for $q \neq 1$,

$$
H_q(P_n) = \frac{-1}{q-1} \log_2 \sum_{k \in K_n} \mu^q[B_{\delta_n}(k)].
$$

The dimensions are then defined by scaling by the box widths.

2.3.1 Definition

Let $K_n = \{k : \mu[B_{\delta_n}(k)] > 0\}$. Define $\widetilde{\theta}(q)$ as

$$\widetilde{\theta}(q) = \lim_{n \to \infty} \frac{\log \sum_{k \in K_n} \mu^q[B_{\delta_n}(k)]}{\log \delta_n} \qquad -\infty < q < \infty, \qquad (2.2)$$

if the limit exists. Note that $-\infty$ is allowed as a limiting value.   □

Let $k(x)$ denote the index of the box that contains the point x, then $B_{\delta_n}(k(x))$ is the box that contains x. To avoid notation becoming clumsy, $B_{\delta_n}(k)$ will be interpreted as the kth box, and $B_{\delta_n}(x)$ as the box that contains the point x. It can be seen that the summation term in Equation 2.2 is just $E\left[\mu^{q-1}[B_{\delta_n}(X)]\right]$, where the expectation is taken with respect to the probability measure μ.

2.3.2 Definition

The *Generalised Rényi Lattice Dimensions*, $\widetilde{D}_q$, are defined as

$$\widetilde{D}_q = \begin{cases} \displaystyle\lim_{n \to \infty} \frac{\sum_{k \in K_n} \mu[B_{\delta_n}(k)] \log \mu[B_{\delta_n}(k)]}{\log \delta_n} & q = 1 \\[2em] \dfrac{\widetilde{\theta}(q)}{q-1} & q \neq 1 \end{cases} \qquad (2.3)$$

when the limit exists for $q = 1$ and whenever $\widetilde{\theta}(q)$ exists for $q \neq 1$.   □

Note that $\widetilde{\theta}(1) = 0$. Further, if the box counting dimension (Definition A.3.1) of the support of μ exists, then it is $\widetilde{D}_0$, where

$$\widetilde{D}_0 = \lim_{n \to \infty} \frac{\log \# K_n}{-\log \delta_n}$$

and $\# K_n$ is the cardinality of K_n, i.e., the number of boxes with positive μ measure.

So far, we have interpreted $\widetilde{\theta}(q)$ as a global average or a measure of information as in information theory. However, there is a third interpretation which we will appeal to in Chapter 4, where it will be interpreted as a rescaled cumulant generating function.

2.3.3 Theorem (Beck, 1990)

The following hold for arbitrary probability measures:

1. $\widetilde{D}_r \leq \widetilde{D}_q$ for any $r > q$; $q, r \in \mathbb{R}$.

2.
$$\frac{r}{r-1} \frac{q-1}{q} \widetilde{D}_q \leq \widetilde{D}_r \leq \widetilde{D}_q$$

   for $r > q > 1$ or $0 > r > q$.   □

Beck (1990) refers to a measure μ as having *minimum uniformity* if

$$\frac{q-1}{q}\,\widetilde{D}_q = \frac{r-1}{r}\,\widetilde{D}_r = \text{constant},$$

or *maximum uniformity* if $\widetilde{D}_q = \widetilde{D}_r = \text{constant}$.

2.3.4 Example

Consider the Cantor measure as in Examples 1.2.1 and 2.2.1. When the measure is not allocated uniformly over the Cantor set, e.g., $p_0 = 1 - p_2 = \frac{1}{3}$ as in Example 1.2.1, it can be seen that $\widetilde{\theta}(q) = -\log_3(p_0^q + p_2^q)$, and so $\widetilde{D}_q = \widetilde{\theta}(q)/(q-1)$.

In the case where the measure is allocated uniformly over the Cantor set, i.e., $p_0 = 1 - p_2 = \frac{1}{2}$, then $\widetilde{\theta}(q) = (q-1)\log_3 2$, and so $\widetilde{D}_q = \log_3 2$ for all q. This is the case of maximum uniformity referred to in Theorem 2.3.3, and $\widetilde{D}_q$ is simply the dimension of the Cantor set. $\square$

2.4 Generalised Rényi Point Centred Dimensions

The second multifractal formalism that we study is based on coverings by spheres with centres within the support of the measure μ. Part of the reason for such a construction was the desire for a more efficient algorithm with which to estimate fractal dimensions. We very briefly outline the general argument here, but return to a more complete discussion of estimation in Part III of the book.

If we were to estimate the box counting dimension of some observed process, we could cover it with a lattice system of boxes of width δ_n, and count the number of boxes N_n that are occupied. This would be done for a sequence of smaller and smaller widths $\{\delta_n\}$. One would then plot $\log N_n$ versus $\log \delta_n$. If δ_n is sufficiently small, then this plot should be a straight line, whose slope is an estimate of the box counting dimension. However, there is much wasted effort in this algorithm, particularly as δ_n gets quite small, because most boxes are never visited by the process. It should also be noted that estimating the dimension of the support of μ, based on observed data, is extremely difficult, because parts of the support may be very rarely visited by the process.

Grassberger and Procaccia (1983a, b, c) suggested an alternative method. Given a sequence of random locations $X_1, X_2, \cdots$, estimate the probability distribution of the interpoint distances, and then estimate the powerlaw exponent ('fractal' dimension) of this probability distribution. That is, take many pairs of independent samples of points, and estimate $\Pr\{\|X_1 - X_2\| \leq \delta\}$ as a function of δ. As in the box counting case, plot $\log \Pr\{\|X_1 - X_2\| \leq \delta\}$ versus $\log \delta$ and estimate the slope of the line. This was then referred to as the *correlation dimension*. However, note that

$$\Pr\{\|X_1 - X_2\| \leq \delta\} = \int \mu[S_\delta(x)]\mu(dx) = \mathrm{E}[\mu[S_\delta(X)]], \qquad (2.4)$$

where $S_\delta(x)$ is a closed sphere of radius δ centred at x, and the expectation is taken with respect to the probability measure μ. So, as in the lattice case, we are estimating the powerlaw exponent of the first order moment, but instead of all spheres having equal weight as the boxes in the lattice case, they are weighted roughly according to the probability that the process visits that part of the space.

In general, we consider the $(q-1)$th order moment

$$\mathrm{E}\left[\mu^{q-1}[S_\delta(X)]\right] = \int \mu^{q-1}[S_\delta(x)]\mu(dx).$$

For given values of q, the Rényi point centred dimensions describe the powerlaw behaviour of $\left\{\mathrm{E}\left[\mu^{q-1}[S_\delta(X)]\right]\right\}^{q-1}$ as $\delta \to 0$. Cutler (1991) described these dimensions using properties of $\mathcal{L}^q$ norms.

2.4.1 Definition

Let $-\infty < q < \infty$ and $\mathcal{X}_\delta = \{x \in \mathcal{X} : \mu[S_\delta(x)] > 0\}$. Then the *point centred correlation exponents* are defined as

$$\theta(q) = \lim_{\delta \to 0} \frac{\log\left[\int_{\mathcal{X}_\delta} \mu^{q-1}[S_\delta(x)]\mu(dx)\right]}{\log \delta} \tag{2.5}$$

given that the limit exists. Note that $\theta(q) = -\infty$ is allowed as a limiting value. □

Upper and lower limits are sometimes analysed when the limit in Equation 2.5 fails to exist. Note that if $\theta(q) = -\infty$ for some $q < 0$, then $\mathrm{E}[\mu^q[S_\delta(X)]] \to \infty$ as $\delta \to 0$ faster than a powerlaw rate (e.g., exponential).

2.4.2 Definition

The *Generalised Rényi Point Centred Dimensions* are denoted by D_q, where

$$D_q = \begin{cases} \dfrac{\theta(q)}{q-1} & q \neq 1 \\[2ex] \lim\limits_{\delta \to 0} \dfrac{\int_{\mathcal{X}_\delta} \log \mu[S_\delta(x)]\mu(dx)}{\log \delta} & q = 1, \end{cases} \tag{2.6}$$

when the limit exists for $q = 1$ and whenever $\theta(q)$ exists for $q \neq 1$.   □

While part of Grassberger & Procaccia's (1983a, b, c) motivation for calculating D_2 was to use a more efficient algorithm than that used to calculate $\tilde{D}_0$, they are both describing quite different characteristics of the observed process. The box counting dimension of the support of μ describes the geometric dimension or size of the support, while the Rényi dimensions D_q (and $\tilde{D}_q$, $q \neq 0$) describe the non-uniformity of the measure.

2.4.3 Note (Cutler, 1991, Page 662)

If μ has a bounded Radon-Nikodym derivative with respect to the uniform measure across its support (i.e., $\mu(E) = \int_E g(x)dx$ where $|g(x)| \le K$), then D_q will be constant for all $q > 0$. However, when μ has singularities, the Rényi dimensions are not the same, and reflect the nature of the singularities. In this situation, D_q $(q \ne 1)$ is a decreasing function of q. $\qquad\qquad\qquad\qquad\qquad\qquad\qquad$ □

This note is the analogue to Theorem 2.3.3 in the lattice case.

2.4.4 Example

Let μ be the uniform measure on $[0, 1]$. Then for $0 < \delta < \frac{1}{2}$,

$$
\mu[S_\delta(x)] = \begin{cases} \delta + x & x < \delta \\ 2\delta & \delta \le x \le 1 - \delta \\ \delta + 1 - x & x > 1 - \delta, \end{cases}
$$

and hence for $q \ne 0$ and $0 < \delta < \frac{1}{2}$

$$
\mathrm{E}\left[\mu^{q-1}[S_\delta(X)]\right]
$$
$$
= \int_0^1 \mu^{q-1}[S_\delta(x)]\mu(dx)
$$
$$
= \int_0^\delta (\delta + x)^{q-1}dx + \int_\delta^{1-\delta} (2\delta)^{q-1}dx + \int_{1-\delta}^1 (\delta + 1 - x)^{q-1}dx
$$
$$
= (2\delta)^{q-1}\left[1 + \frac{2\delta}{q}\left(2 - q - 2^{1-q}\right)\right].
$$

Note that for $q < 0$, $\mathrm{E}\left[\mu^{q-1}[S_\delta(X)]\right] \to \infty$ as $\delta \to 0$. $\theta(q)$ describes the *powerlaw behaviour* of $\mathrm{E}[\mu^q[S_\delta(X)]]$ as $\delta \to 0$. It can be seen that $\theta(q) = \lim_{\delta \to 0} \theta_\delta(q) = q - 1$, therefore $D_q = 1$ for all $q \ne 0$, consistent with Note 2.4.3. $\qquad\qquad\qquad\qquad\qquad\qquad\qquad\qquad\qquad\qquad$ □

The probability distribution of interpoint distances described by Equation 2.4 can be extended to a sample of q independent points. This relationship forms the basis of the method we use to estimate D_q in Part III of the book.

2.4.5 Theorem - Distribution of qth Order Interpoint Distance

Let $X_1, X_2, \cdots, X_q$ be a sample of independent random variables drawn from the probability distribution μ, and define Y as

$$
Y = \max\{\|X_1 - X_q\|, \|X_2 - X_q\|, \cdots, \|X_{q-1} - X_q\|\}.
$$

Then for $q = 2, 3, 4, \cdots$

$$
\int \mu[S_\delta(x)]^{q-1}\mu(dx) = \Pr\{Y \le \delta\}.
$$

Proof. Let $1(A)$ be one if A is true and zero otherwise, then

$$\mu[S_\delta(x)] = \int 1(\|x_1 - x\| \le \delta)\,\mu(dx_1) = \Pr\{\|X_1 - x\| \le \delta\}.$$

Further,

$$\mu^{q-1}[S_\delta(x)]$$
$$= \int \cdots \int 1(\|x_1 - x\| \le \delta) \cdots 1(\|x_{q-1} - x\| \le \delta)\,\mu(dx_1) \cdots \mu(dx_{q-1})$$
$$= \int \cdots \int 1(\max\{\|x_1 - x\|, \cdots, \|x_{q-1} - x\|\} \le \delta)$$
$$\times \mu(dx_1) \cdots \mu(dx_{q-1}),$$

and so it follows that

$$\int \mu^{q-1}[S_\delta(x)]\mu(dx)$$
$$= \int \int \cdots \int 1(\max\{\|x_1 - x_q\|, \cdots, \|x_{q-1} - x_q\|\} \le \delta)$$
$$\times \mu(dx_1) \cdots \mu(dx_{q-1})\mu(dx_q)$$
$$= \Pr\{Y \le \delta\}.$$

$\square$

2.5 Multifractal Spectrum and Formalism

Here we describe *local* behaviour of the measure μ, that is, the behaviour on individual lattices or spheres.

2.5.1 Definition

Let

$$K_n(y, \epsilon) = \left\{ k : y - \epsilon < \frac{\log \mu[B_{\delta_n}(k)]}{\log \delta_n} \le y + \epsilon \right\}.$$

The *Multifractal Spectrum* in the lattice case, denoted by $\widetilde{f}(y)$, is defined to be

$$\widetilde{f}(y) = \lim_{\epsilon \to 0} \lim_{n \to \infty} \frac{\log \#K_n(y, \epsilon)}{-\log \delta_n} \tag{2.7}$$

for $y > 0$, allowing for $\widetilde{f}(y) = -\infty$ when $\#K_n(y, \epsilon) = 0$. $\square$

The function $\widetilde{f}(y)$ can be expressed differently. Let $B_{\delta_n}(k(x))$ denote the box that contains the point x; i.e., x is contained in the $k(x)$th box. To avoid notation becoming too clumsy, we will simply denote this as $B_{\delta_n}(x)$, if in the given context, it cannot be confused with $B_{\delta_n}(k)$ where $k \in K_n$. Note that $\mu[B_{\delta_n}(x')]$ is constant for all $x' \in B_{\delta_n}(x)$, therefore,

$$\#K_n(y, \epsilon) = \int_{\widetilde{F}_n(y,\epsilon)} \frac{\mu(dx)}{\mu[B_{\delta_n}(x)]},$$

where

$$\widetilde{F}_n(y, \epsilon) = \left\{ x \in \mathcal{X} : y - \epsilon < \frac{\log \mu[B_{\delta_n}(x)]}{\log \delta_n} \leq y + \epsilon \right\}.$$

Therefore, $\widetilde{f}(y)$ can also be expressed as

$$\widetilde{f}(y) = \lim_{\epsilon \to 0} \lim_{n \to \infty} \frac{1}{-\log \delta_n} \log \int_{\widetilde{F}_n(y,\epsilon)} \frac{\mu(dx)}{\mu[B_{\delta_n}(x)]}. \tag{2.8}$$

Equation 2.8 provides the format for the analogue to Equation 2.7 for the multifractal spectrum in the point centred situation.

2.5.2 Definition

Let

$$F_\delta(y, \epsilon) = \left\{ x \in \mathcal{X} : y - \epsilon < \frac{\log \mu[S_\delta(x)]}{\log \delta} \leq y + \epsilon \right\}.$$

The *Multifractal Spectrum* in the point centred case, denoted by $f(y)$, is defined as

$$f(y) = \lim_{\epsilon \to 0} \lim_{\delta \to 0} \frac{1}{-\log \delta} \log \int_{F_\delta(y,\epsilon)} \frac{\mu(dx)}{\mu[S_\delta(x)]} \tag{2.9}$$

for $y > 0$, with the convention that $\log 0 = -\infty$, hence allowing $f(y) = -\infty$ when the integral is zero.   □

2.5.3 Note

Note the similarity between Equations 2.8 and 2.9 when $\mu[S_\delta(x)] \approx$ const for all $x \in F_\delta(y, \epsilon)$. Then

$$\int_{F_\delta(y,\epsilon)} \frac{\mu(dx)}{\mu[S_\delta(x)]} \approx \frac{\mu[F_\delta(y, \epsilon)]}{\text{const}},$$

which is roughly the number of spheres needed to cover $F_\delta(y, \epsilon)$.   □

2.5.4 Definition

1. The measure μ will be said to be a *multifractal measure* satisfying the lattice formalism in a *weak sense* if $\widetilde{\theta}(q)$ and $\widetilde{f}(y)$, as defined by Equations 2.2 and 2.7 respectively, exist and satisfy the Legendre transform pair

$$\widetilde{\theta}(q) = \inf_y \left\{ qy - \widetilde{f}(y) \right\}$$

and

$$\tilde{f}(y) = \inf_q \left\{ qy - \tilde{\theta}(q) \right\}.$$

2. The measure μ will be said to be a *multifractal measure* satisfying the lattice formalism in a *strong sense* if it does in a weak sense, and for $\tilde{f}(y) > 0$,

$$\dim_H \tilde{F}(y) = \tilde{f}(y),$$

where $\tilde{F}(y) = \{x \in \mathcal{X} : \lim_{n \to \infty} Y_n(x) = y\}$, $\dim_H$ is the Hausdorff dimension, and

$$Y_n(x) = \frac{\log \mu[B_{\delta_n}(x)]}{\log \delta_n}. \tag{2.10}$$

□

It may be tempting to argue that $\tilde{f}(y) = \dim_B \tilde{F}(y)$ since $\tilde{f}(y)$ involves some sort of box covering. This is not necessarily the case, as the limit in Equation 2.7 not only determines the nature of the set in question but also the size of the lattice covering. In the case of $\dim_B \tilde{F}(y)$, the set $\tilde{F}(y)$ is defined quite separately to the size and number of covers used to determine the box dimension. Further, it is often the case that sets of the form of $\tilde{F}(y)$ are dense, in which case $\dim_B \tilde{F}(y) = \dim_B \text{supp}(\mu)$.

Note that $\tilde{f}(y)$ is sometimes referred to as the *coarse grained* multifractal spectrum, and $\dim_H \tilde{F}(y)$ or $\dim_P \tilde{F}(y)$ as the *fine grained* multifractal spectrum, where $\dim_H$ and $\dim_P$ denote the Hausdorff (Definition A.2.3) and packing (Definition A.4.2) dimensions respectively.

2.5.5 Definition

1. The measure μ will be said to be a *multifractal measure* satisfying the point centred formalism in a *weak sense* if $\theta(q)$ and $f(y)$, as defined by Equations 2.5 and 2.9 respectively, exist and satisfy the Legendre transform pair

$$\theta(q) = \inf_y \left\{ qy - f(y) \right\}$$

and

$$f(y) = \inf_q \left\{ qy - \theta(q) \right\}.$$

2. The measure μ will be said to be a *multifractal measure* satisfying the point centred formalism in a *strong sense* if it does in a weak sense, and for $f(y) > 0$,

$$\dim_H F(y) = f(y),$$

where $F(y) = \{x \in \mathcal{X} : \lim_{\delta \to 0} Y_\delta(x) = y\}$, $\dim_H$ is the Hausdorff dimension, and

$$Y_\delta(x) = \frac{\log \mu[S_\delta(x)]}{\log \delta}.$$ (2.11)

□

Notice the difference between the sets $\widetilde{F}_n(y, \epsilon)$ and $\widetilde{F}(y)$ and also between $F_\delta(y, \epsilon)$ and $F(y)$. The sets $\widetilde{F}_n(y, \epsilon)$ and $F_\delta(y, \epsilon)$ describe the local behaviour in a pre-limiting sense, while $\widetilde{F}(y)$ and $F(y)$ describe the limiting local behaviour.

In a series of recent papers by Mandelbrot & Riedi, they introduce the inverse measure of μ, say μ^*, where the multifractal spectrums are related by the 'inversion formula' $f^*(y) = yf(1/y)$. Further discussion can be found in Riedi & Mandelbrot (1997, 1998) and Mandelbrot & Riedi (1997).

2.6 Review of Related Lattice Based Results

In our definitions of $\widetilde{f}(y)$ and $\widetilde{\theta}(q)$ in the lattice case, the limit has been taken with respect to a predetermined sequence $\{\delta_n\} \to 0$ as $n \to \infty$. However, the limits may not exist for all sequences ($\delta \to \infty$). Another situation is the case where lim inf and lim sup as in Equation 2.2 are not the same. In this section, we review results relating to these two situations.

2.6.1 Alternative Relationship (Falconer, 1990, §17.1)

Falconer defines $\widetilde{\theta}(q)$ differently, as

$$\widetilde{\theta}(q) = \lim_{\delta \to 0} \frac{\log \sum_{k \in K_n} \mu^q[B_\delta(k)]}{\log \delta},$$ (2.12)

i.e., the limit is assumed to exist for all possible sequences. Note in our formulation, assumptions about the nature and existence of $\widetilde{\theta}(q)$ will be made (Chapter 4), and from this, the behaviour of $\widetilde{f}(y)$ will be deduced. Falconer (1990) poses the problem from the opposite direction. For $0 \le y < \infty$, assume that $\widetilde{f}(y)$ in Equation 2.7 exists but with $\delta \to 0$ (i.e., for all sequences). Then Falconer (1990) shows that the limit $\widetilde{\theta}(q)$ given by Equation 2.12 exists, and that $\widetilde{\theta}(q) = \inf_y \{qy - \widetilde{f}(y)\}$. □

However, the limit is sometimes dependent on the sequence δ_n, as shown by Riedi (1995) in the following example.

2.6.2 Example (Riedi, 1995, §4.1)

Let μ be the Cantor measure as in Example 2.3.4. Then when $\delta_n = 3^{-n}$, $\widetilde{\theta}(q) = -\log_3(p_0^q + p_2^q)$ for all $q \in \mathbb{R}$. However, Riedi (1995) shows that

$$\liminf_{\delta \to 0} \frac{\log \sum_k' \mu^q[B_\delta(k)]}{\log \delta} = -\infty,$$ (2.13)

when $q < 0$, and where the prime on the summation indicates that it is to be taken over those boxes with positive μ measure.

This is done as follows. For every $n \in \mathbb{N}$, one can select a $h_n \in \mathbb{N}$ satisfying both $p_2^{h_n} \leq (\frac{1}{2}3^{-n})^n$ and $h_n \geq n+1$. Then let $\delta_n = 3^{-n}(1-3^{-h_n})$. A δ_n-mesh is applied to $[0,1]$, pivoting as usual at the origin. By construction, the last box, say $B_{\delta_n}(*)$, only intersects with $[0,1]$ a very small amount, i.e.,

$$B_{\delta_n}(*) \cap [0,1] = \left[1 - 3^{-h_n}, 1\right],$$

which is the last triadic interval of width 3^{-h_n}. Therefore,

$$\mu[B_{\delta_n}(*)] = p_2^{h_n} \leq \left(\frac{3^{-n}}{2}\right)^n < \delta_n^n,$$

since $1 - 3^{-h_n} > \frac{1}{2}$. Thus, when $q < 0$, $\sum_k' \mu^q[B_{\delta_n}(k)] > \delta_n^{nq}$. Hence

$$\frac{\log \sum_k' \mu^q[B_{\delta_n}(k)]}{\log \delta_n} < nq$$

and, hence, Equation 2.13.   □

2.6.3 Overlapping Boxes (Riedi, 1995)

Riedi defines a system of overlapping boxes where $\delta \to 0$ for all sequences, so that the anomaly of Example 2.6.2 is avoided. Denote the kth box in the δ-mesh as $B_\delta(k)$, but let $\delta \to 0$ continuously rather than discretely as in Definition 2.3.1. In particular,

$$B_\delta(k) = [k_1\delta, (k_1+1)\delta) \times \cdots \times [k_d\delta, (k_d+1)\delta)$$

where $(k_1, \cdots, k_d)$ are the grid coordinates of the kth box. Let $K_\delta = \{k : \mu[B_\delta(k)] > 0\}$. Riedi (1995) defines a system with the same number of boxes, $\#K_\delta$, but three times the width, hence overlapping. Denote the kth box as $\overline{B}_\delta(k)$ where

$$\overline{B}_\delta(k) = [(k_1-1)\delta, (k_1+2)\delta) \times \cdots \times [(k_d-1)\delta, (k_d+2)\delta)$$

and $k \in K_\delta$; i.e., k takes the same values as the lattice index in $B_\delta(k)$.

He then defines the $\widetilde{\theta}$-function as

$$\widetilde{\theta}_R(q) = \liminf_{\delta \to 0} \frac{1}{\log \delta} \log \sum_{k \in K_\delta} \mu^q\left[\overline{B}_\delta(k)\right].$$

We add the subscript R to distinguish it from $\widetilde{\theta}(q)$. If $\liminf_{\delta \to 0}$ can be replaced by $\lim_{\delta \to 0}$ for a particular q, then $\widetilde{\theta}_R(q)$ is said to be *grid regular*. Similarly,

$$\widetilde{f}_R(y) = \lim_{\epsilon \to 0} \limsup_{\delta \to 0} \frac{\log \#K_\delta(y, \epsilon)}{-\log \delta},$$

where

$$K_\delta(y, \epsilon) = \left\{ k \in K_\delta : y - \epsilon < \frac{\log \mu[\overline{B}_\delta(k)]}{\log \delta} \leq y + \epsilon \right\}.$$

Again, $\widetilde{f}_R(y)$ is said to be *grid regular* if $\lim \sup_{\delta \to 0}$ can be replaced by $\lim_{\delta \to 0}$. He then shows that

$$\widetilde{\theta}_R(q) = \inf_y \left\{ qy - \widetilde{f}_R(y) \right\} \qquad q \neq 0.$$

Also, if $\widetilde{\theta}_R(q)$ is grid regular, differentiable and convex on $\mathbb{R}$, then

$$\widetilde{f}_R(y) = \lim_{\epsilon \to 0} \lim_{\delta \to 0} \frac{\log \# K_\delta(y, \epsilon)}{-\log \delta} = \inf_q \left\{ qy - \widetilde{\theta}_R(q) \right\}.$$

Further, if $\widetilde{\theta}_R(q)$ is differentiable at $q \neq 0$ and if $y_q = \widetilde{\theta}'_R(q)$, then

$$\widetilde{\theta}_R(q) = q y_q - \widetilde{f}_R(y_q).$$

2.6.4 Upper and Lower Bounds

Brown et al. (1992) analyse lattice coverings of the unit interval. They investigated conditions under which the Hausdorff and packing dimensions of the following sets can be determined:

$$\left\{ x \in [0, 1) : \liminf_{n \to \infty} Y_n(x) \leq y \right\}, \qquad \left\{ x \in \mathrm{supp}(\mu) : \liminf_{n \to \infty} Y_n(x) \geq y \right\},$$

$$\left\{ x \in [0, 1) : \limsup_{n \to \infty} Y_n(x) \leq y \right\}, \text{ and } \left\{ x \in \mathrm{supp}(\mu) : \limsup_{n \to \infty} Y_n(x) \geq y \right\},$$

where $Y_n(x)$ is given by Equation 2.10. $\qquad\qquad\qquad\qquad\qquad\qquad\qquad\square$

2.7 Review of Related Point Centred Results

In this section, we briefly review some results relating to point centred constructions. Generally, these results describe relationships between the lower and upper limits of the local behaviour of the measure and the Hausdorff and packing dimensions, respectively.

2.7.1 Theorem (Young, 1982; Pesin, 1993)

Let $S_\delta(x)$ be a closed sphere of radius δ centred at x. If μ is a probability measure on $\mathcal{B}(\mathcal{X})$ and if

$$\lim_{\delta \to 0} \frac{\log \mu[S_\delta(x)]}{\log \delta}$$

exists and is constant for μ almost all x, then the limit is given by σ where

$$\sigma = \inf\{\dim_H(E) : \mu(E) = 1\}. \qquad (2.14)$$

□

Note that E is generally not closed, and therefore the Hausdorff dimension of the support of μ may be strictly larger than σ.

In a series of papers, Cutler (1986, 1991) and Cutler & Dawson (1989) have given a more detailed description of the local behaviour of the measure μ and its relationship with the Hausdorff and packing dimensions. In Definition 2.5.5, we simply defined $F(y)$ assuming that $\lim_{\delta \to 0} Y_\delta(x)$ existed. Cutler considers the individual lim inf and lim sup limits, defined below, which appear to be related to the Hausdorff and packing dimensions respectively. These are then related to *dimension distributions*. These ideas were also discussed earlier by Gács (1973).

2.7.2 Definitions (Cutler, 1991)

1. Define the probability measures μ_H and μ_P as

$$\mu_H([0, y]) = \sup\{\mu(E) : \dim_H(E) \le y\}$$

and

$$\mu_P([0, y]) = \sup\{\mu(E) : \dim_P(E) \le y\}$$

respectively.

2. The measure μ is said to be *dimension regular* if $\mu_H = \mu_P$.

3. The measure μ is of *exact Hausdorff (respectively, exact packing) dimension* if there exists $y_1 \ge 0$ such that $\mu_H = \delta_{y_1}$ (respectively, $\mu_P = \delta_{y_1}$) where δ_{y_1} is the unit mass at y_1.

4. Define the local mappings $Y_{\inf} : \mathcal{X} \to [0, \infty]$ and $Y_{\sup} : \mathcal{X} \to [0, \infty]$ as

$$Y_{\inf}(x) = \liminf_{\delta \to 0} \frac{\log \mu[S_\delta(x)]}{\log \delta}$$

and

$$Y_{\sup}(x) = \limsup_{\delta \to 0} \frac{\log \mu[S_\delta(x)]}{\log \delta}.$$

□

The dimension distributions μ_H and μ_P are defined above. The following result states that $\lim_{\delta \to 0} Y_\delta(x)$, which determines the local behaviour of μ, exits μ-a.s. iff the two dimension distributions are the same. As such there is an explicit relationship between the Hausdorff dimension and lim inf and the packing dimension and lim sup.

2.7.3 Theorem (Cutler, 1991)

$Y_{\text{inf}}(x) = Y_{\text{sup}}(x)$ μ-a.s. if and only if μ is dimension regular. Further, μ is of exact Hausdorff (respectively, packing) dimension y_1 if and only if $Y_{\text{inf}} = y_1$ μ-a.s. (respectively, $Y_{\text{sup}} = y_1$ μ-a.s.). □

2.7.4 Corollary (Cutler, 1991)

The limit

$$\lim_{\delta \to 0} \frac{\log \mu[S_\delta(x)]}{\log \delta} = y_1 = \text{constant}$$

μ-a.s. if and only if μ is dimension regular and of exact dimension y_1. In this case $y_1 = \sigma$, where σ is given by Equation 2.14. □

2.7.5 Theorem (Cutler, 1991, Page 657)

Define the sets

$$G_{\text{inf}}(y) = \{x : Y_{\text{inf}}(x) \leq y\}$$

and

$$G_{\text{sup}}(y) = \{x : Y_{\text{sup}}(x) \leq y\}.$$

The following relations hold:

1. $\dim_H G_{\text{inf}}(y) \leq y$ and $\dim_P G_{\text{sup}}(y) \leq y$, and

2. $\mu_H([0, y]) = \mu[G_{\text{inf}}(y)]$ and $\mu_P([0, y]) = \mu[G_{\text{sup}}(y)]$. □

Alternatively, one might consider $G'_{\text{sup}}(y) = \{x : Y_{\text{sup}}(x) \geq y\}$. Then $F(y) = G_{\text{inf}}(y) \cap G'_{\text{sup}}(y)$.

2.7.6 Average Local Behaviour

If a point X is chosen randomly with respect to μ, then $Y_{\text{inf}}(X)$ and $Y_{\text{sup}}(X)$ may be regarded as random variables with distributions μ_H and μ_P, respectively. Define the average Hausdorff and packing local limits respectively by

$$E[Y_{\text{inf}}(X)] = \int Y_{\text{inf}}(x)\mu(dx) = \int_0^\infty y\,\mu_H(dy),$$

and

$$E[Y_{\text{sup}}(X)] = \int Y_{\text{sup}}(x)\mu(dx) = \int_0^\infty y\,\mu_P(dy).$$

$E[Y_{\text{inf}}(X)] = E[Y_{\text{sup}}(X)]$ if and only if μ is dimension regular. If μ is also of exact dimension σ, then $\sigma = E[Y_{\text{inf}}(X)] = E[Y_{\text{sup}}(X)]$. □

The two means described above have a similar interpretation to D_1, and in many situations will be the same. If the measure is dimension regular, then the only remaining difference is the interchanging of the integration and the limit in δ.

Cutler (1991) also defines the $\mathcal{L}^q$ norm of $\mu[S_\delta(x)]$ and relates it to the Rényi dimensions D_q. An analogous result to that of Theorem 2.3.3 in the lattice case follows.

2.7.7 Theorem (Cutler, 1991, Page 659)

Let μ be a probability measure on the Borel sets, $\mathcal{B}(\mathcal{X})$, of a compact set $\mathcal{X} \subseteq \mathbb{R}^d$. Also let $\theta^-(q)$ and $\theta^+(q)$ be the same as $\theta(q)$ in Equation 2.5 except where $\lim_{\delta \to 0}$ is replaced by $\liminf_{\delta \to 0}$ and $\limsup_{\delta \to 0}$, respectively. Then the following hold.

1. For $q \neq 1$,

$$\left\{ \frac{1}{q-1} \theta^-(q) \right\} \quad \text{and} \quad \left\{ \frac{1}{q-1} \theta^+(q) \right\}$$

are decreasing as functions of q.

2. For all r and q such that $r < 1 < q$,

$$\frac{1}{r-1} \theta^-(r) \geq \mathrm{E}[Y_{\inf}(X)] \quad \text{and} \quad \mathrm{E}[Y_{\sup}(X)] \geq \frac{1}{q-1} \theta^+(q).$$

3. If μ is dimension regular then, whenever $r < 1 < q$,

$$\frac{1}{r-1} \theta^-(r) \geq \mathrm{E}[Y_{\inf}(X)] = \mathrm{E}[Y_{\sup}(X)] \geq \frac{1}{q-1} \theta^+(q).$$

$\square$

Cutler (1995) discusses further results of dual representations of the Hausdorff and packing dimensions in terms of the lower and upper local mappings $Y_{\inf}(x)$ and $Y_{\sup}(x)$, respectively. This is done in both a 'strong' and a 'weak' sense.

Pesin (1993) and Olsen (1994, pages 24–26) discuss various alternative definitions of generalised dimensions and their relationships.

CHAPTER 3

The Multinomial Measure

3.1 Introduction

The multinomial measure is a generalised version of the Cantor measure discussed in Example 1.2.1. For this example we derive expressions for both $\widetilde{\theta}(q)$ as in Equation 2.2, and the multifractal spectrum $\widetilde{f}(y)$ as in Equation 2.7. We then show that $\widetilde{\theta}(q)$ and $\widetilde{f}(y)$ are indeed related via the Legendre transform. This is done relatively simply using Lagrange multipliers. The Legendre transform also appears in the theory of large deviations, and in Part II of the book we show that this theory can be used to derive sufficient conditions for the Legendre transform relationships to hold in a more general setting.

A multinomial measure is constructed on the unit interval by repeatedly dividing the interval and reallocating mass. The unit interval is initially divided into b subintervals of equal length, where $b \geq 2$. Each of these subintervals is then divided into b further subintervals, etc. Each subinterval at the nth iteration can be characterised uniquely by a b-adic rational number of length n. Further, the value of the measure on a subinterval can also be defined in terms of the mix of digits in the respective b-adic number. This example has been discussed by Mandelbrot (1989).

3.1.1 Construction of the Measure

Let $\Omega = \{0, 1, \cdots, (b-1)\}$ where $b \in \{2, 3, \cdots\}$ and is fixed. Assign to each element $\omega \in \Omega$ a probability p_ω such that

$$\sum_{\omega=0}^{b-1} p_\omega = 1.$$

Also let $\Omega_0 = \{\omega \in \Omega : p_\omega > 0\}$, $s = \#\Omega_0$, and Ω^n and Ω_0^n be the nth cross products of Ω and Ω_0 respectively.

Further consider an iterative scheme on the unit interval $[0, 1]$, and the construction of a probability measure μ. At the first iteration, divide $[0, 1]$ into b non-overlapping closed subintervals. The total mass (or probability) is then allocated to each of the subintervals according to the proportions $p_0, \cdots, p_{b-1}$. At each subsequent iteration, each subinterval is divided into b further subintervals. Again, the total mass contained in an interval before being divided is allocated to each of the subintervals according to the proportions $p_0, \cdots, p_{b-1}$. After n

iterations, $[0, 1]$ has been divided into b^n subintervals, each of equal length; of which s^n contain non-zero mass.

By considering the sequences $\omega \in \Omega^n$ as b-adic rational numbers on the interval $[0, 1]$, each sequence represents one of the b^n subintervals (of width $\delta_n = b^{-n}$); in that, the base b expansion of all numbers in the interior of a particular subinterval will have the same first n digits. Subintervals at the nth stage will be represented as $J_n(\omega)$, where

$$J_n(\omega) = \left[\sum_{j=1}^{n} \frac{\omega_j}{b^j}, \ b^{-n} + \sum_{j=1}^{n} \frac{\omega_j}{b^j} \right],$$

and $\omega_j \in \Omega$. The probability measure, μ, attributable to a particular interval is then

$$
\begin{aligned}
\mu[J_n(\omega)] &= p_{\omega_1} \cdots p_{\omega_n} \\
&= p_0^{n\Phi_0(\omega)} p_1^{n\Phi_1(\omega)} \cdots p_{b-1}^{n\Phi_{b-1}(\omega)},
\end{aligned}
\tag{3.1}
$$

where $\Phi_\alpha(\omega)$ is the fraction of digits that equal α in the sequence $\omega \in \Omega^n$. Therefore, the nth cross product of Ω_0 is $\Omega_0^n = \{\omega \in \Omega^n : \mu[J_n(\omega)] > 0\}$.

3.1.2 Example

An example of a multinomial measure is the Cantor measure as in Example 1.2.1. For that particular case $b = 3$, $p_0 = \frac{1}{3}$, $p_1 = 0$, and $p_2 = \frac{2}{3}$; thus $s = 2$. □

3.2 Local Behaviour

A probability measure was constructed on a set with zero Lebesgue measure in Example 1.2.1. Such a measure does not have a probability density function. An alternative way to characterise such a measure is to describe what we will refer to as its local behaviour.

For a given n, consider those intervals $J_n(\omega)$ with non-zero measure, i.e., $\omega \in \Omega_0^n$. For these intervals, their local behaviour is described by $Y_n(\omega)$, where

$$
\begin{aligned}
Y_n(\omega) &= \frac{\log \mu[J_n(\omega)]}{\log \delta_n} \\
&= \frac{-1}{n} \log_b \left[\prod_{\alpha \in \Omega_0^n} p_\alpha^{n\Phi_\alpha(\omega)} \right] \\
&= - \sum_{\alpha \in \Omega_0} \Phi_\alpha(\omega) \log_b p_\alpha.
\end{aligned}
\tag{3.2}
$$

We wish to treat Y_n as a discrete random variable. That is, if we were to randomly select an interval $J_n(\omega)$ from those that have positive measure, each with equal probability, what is the probability that Y_n is equal to y, i.e., $\Pr\{Y_n = y\}$?

The problem can be simplified by noticing its *discrete* nature. For example, let $b = 2$ and $n = 4$, then $\Phi_\alpha(\omega) = 0, \frac{1}{4}, \frac{1}{2}, \frac{3}{4}$ or 1 ($\alpha = 0, 1$). The above probability distribution then can be reduced to an expression in terms of the number of subintervals where Φ_0 and Φ_1 take on particular values. For example, if $\Phi_0 = \Phi_1 = \frac{1}{2}$, then there are 6 possible b-adic rational expansions of length $n = 4$ with these values:

$$0.0011 \quad 0.0101 \quad 0.0110$$
$$0.1010 \quad 0.1100 \quad 0.1001.$$

Define the set of $(b \times 1)$ vectors $\mathbf{\Psi}(y)$ as

$$\mathbf{\Psi}(y) = \left\{ (\phi_0, \cdots, \phi_{b-1}) : y = - \sum_{\omega \in \Omega_0} \phi_\omega \log_b p_\omega \text{ and } \sum_{\omega \in \Omega_0} \phi_\omega = 1 \right\}, \quad (3.3)$$

and a subset $\mathbf{\Psi}_n(y)$ as

$$\mathbf{\Psi}_n(y) = \{ (\phi_0, \cdots, \phi_{b-1}) \in \mathbf{\Psi}(y) : n\phi_\omega \text{ is a non} -\text{ve integer } \forall \omega \in \Omega_0 \}.$$

Note that if $s = 2$ and $p_{\omega_1} \neq p_{\omega_2}$, where $\Omega_0 = \{\omega_1, \omega_2\}$, then $\mathbf{\Psi}(y)$ contains a *unique* point. Given particular values of $(\phi_0, \cdots, \phi_{b-1}) \in \mathbf{\Psi}_n(y)$, the number of subintervals at the nth step with these values, denoted by $N_n(\phi_0, \cdots, \phi_{b-1})$ is

$$N_n(\phi_0, \cdots, \phi_{b-1}) = \frac{n!}{(n\phi_0)! \cdots (n\phi_{b-1})!}.$$

It therefore follows that

$$\Pr\{Y_n = y\} = \begin{cases} 0 & \text{if } \mathbf{\Psi}_n(y) = \emptyset \\ \dfrac{1}{s^n} \displaystyle\sum_{(\phi_0, \cdots, \phi_{b-1}) \in \mathbf{\Psi}_n(y)} N_n(\phi_0, \cdots, \phi_{b-1}) & \text{if } \mathbf{\Psi}_n(y) \neq \emptyset. \end{cases}$$

Note that $\Pr\{Y_n = y\} \to 0$ unless $y = y_0 = (1/s) \sum_{\omega \in \Omega_0} \log_b p_\omega$. However, the unit interval can be partitioned into various parts according to the powerlaw rates at which $\Pr\{Y_n = y\}$ tends to zero.

3.2.1 Multifractal Spectrum

The multifractal spectrum, denoted by $\widetilde{f}(y)$, describes convergence rates of the probability distribution of Y_n. The number of subintervals at the nth stage with positive μ measure where $Y_n = y$ is $s^n \Pr\{Y_n = y\}$ and the widths of the subintervals are $\delta_n = b^{-n}$ (see Figure 1.2). Hence, in the present example the multifractal spectrum is

$$\widetilde{f}(y) = \lim_{n \to \infty} \frac{\log (s^n \Pr\{Y_n = y\})}{- \log \delta_n} \quad (3.4)$$

$$= \lim_{n \to \infty} \frac{1}{n} \log_b \sum_{(\phi_0, \cdots, \phi_{b-1}) \in \mathbf{\Psi}_n(y)} N_n(\phi_0, \cdots, \phi_{b-1}).$$

Note that

$$
\max_{(\phi_0,\cdots,\phi_{b-1})\in\Psi_n(y)} N_n(\phi_0,\cdots,\phi_{b-1})
$$

$$
\leq \sum_{(\phi_0,\cdots,\phi_{b-1})\in\Psi_n(y)} N_n(\phi_0,\cdots,\phi_{b-1})
$$

$$
\leq (\#\Psi_n(y)) \max_{(\phi_0,\cdots,\phi_{b-1})\in\Psi_n(y)} N_n(\phi_0,\cdots,\phi_{b-1})
$$

and $\#\Psi_n(y) \leq (n+1)^b$. Therefore

$$
\widetilde{f}(y) = \max_{(\phi_0,\cdots,\phi_{b-1})\in\Psi(y)} \lim_{n\to\infty} \frac{1}{n} \log_b N_n(\phi_0,\cdots,\phi_{b-1}).
$$

It follows from Stirling's approximation that

$$
\lim_{n\to\infty} \frac{1}{n} \log_b N_n(\phi_0,\cdots,\phi_{b-1}) = -\sum_{\omega\in\Omega_0} \phi_\omega \log_b \phi_\omega, \tag{3.5}
$$

hence

$$
\widetilde{f}(y) = \max_{(\phi_0,\cdots,\phi_{b-1})\in\Psi(y)} -\sum_{\omega\in\Omega_0} \phi_\omega \log_b \phi_\omega. \tag{3.6}
$$

$\square$

The form of the function $\widetilde{f}(y)$, as in Equation 3.4, is like a box counting dimension in that it is the number of covering boxes required divided by the box width where this width tends to zero. However, it is also different from a box counting dimension, as the nature of the set being covered is also changing as n tends to infinity.

3.3 Global Averaging and Legendre Transforms

An alternative way to describe the measure μ is to use a global average of qth powers of μ, or qth moment. We first define this and then show that the global averages and local behaviour are related by a Legendre transformation.

3.3.1 Global Averaging

In this example, the correlation exponents defined by Equation 2.2 can be expressed as

$$
\widetilde{\theta}(q) = \lim_{n\to\infty} \frac{\log \sum_{\omega\in\Omega_0^n} \mu^q[J_n(\omega)]}{\log \delta_n}, \tag{3.7}
$$

where $q \in \mathbb{R}$ and $\delta_n = b^{-n}$ is the subinterval width. It then follows that

$$
\begin{aligned}
\widetilde{\theta}(q) &= \lim_{n \to \infty} \frac{-1}{n} \log_b \left[\sum_{\omega_1 \in \Omega_0} \cdots \sum_{\omega_n \in \Omega_0} p_{\omega_1}^q \cdots p_{\omega_n}^q \right] \\
&= -\log_b \sum_{\omega \in \Omega_0} p_\omega^q . \qquad\qquad (3.8)
\end{aligned}
$$

$\square$

In the literature, $\widetilde{\theta}(q)$ is often denoted by $\tau(q)$. The reason for using θ is that in Part III of the book, we consider estimation techniques for $\theta(q)$. Using θ instead of τ makes it easier to distinguish between the real value θ, a sample estimate $\widehat{\theta}$, and an estimator Θ.

3.3.2 Example

Consider the case of the multinomial measure with $b = 10$. If we used a small value of b, say two or three, then a relatively large value of one p_ω would necessarily mean smaller values of all other p_ω's. We consider three scenarios: the first is where $p_\omega = 0.1$ for all ω (i.e., $\omega = 0, \cdots, 9$). This will simply produce the uniform distribution. The corresponding plot of $\widetilde{\theta}(q)$ is the straight line with slope one in Figure 3.1. The second scenario is where all values of $p_\omega = 0.111$ except

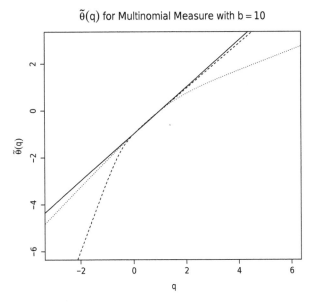

Figure 3.1 *The straight (solid) line represents the case where $p_\omega = 0.1$ for all ω (i.e., $\omega = 0, \cdots, 9$). The dashed line is where $p_\omega = 0.111$ for all ω, except one that is equal to 0.001; and the dotted line is where all p_ω's are equal to 0.07, except one that is equal to 0.37.*

one that is equal to 0.001. This is like the uniform distribution, but it has small areas where there is very small (non-zero) measure. The corresponding graph of $\widetilde{\theta}(q)$ has the line dropping more sharply for $q < 0$ but similar to that of the uniform measure for $q > 0$. The last case is where all p_ω's are equal to 0.07, except one that is equal to 0.37. In this case the plot of $\widetilde{\theta}(q)$ drops away more sharply for $q > 0$, but is fairly similar to that of the uniform measure for $q < 0$.

When all non-zero p_ω's are equal, the measure will be uniformly distributed over the support of μ. In this situation, it follows from Equation 3.8 that $\widetilde{\theta}(q)$ will be a straight line with slope $\widetilde{D}_0$. When the measure is not uniformly distributed over its support, $\widetilde{\theta}(q)$ becomes concave downwards, i.e., the two ends of the line are pulled downwards remaining anchored at $q = 0$, but with the line still increasing as q increases. For $q > 0$ and increasing, the line will be pulled further down from that representing the uniform case if there are areas that contain much greater measure relative to other areas. Conversely, for $q < 0$ and decreasing, the line will be pulled further down from that representing the uniform case if there are areas that contain very small *non-zero* amounts of measure. In summary, the function $\widetilde{\theta}(q)$, therefore, could be interpreted as the extent to which the measure deviates from a measure uniformly distributed on its support. When $q > 0$, it describes the abundance of areas of relatively high measure, and when $q < 0$, it describes areas containing relatively small but *non-zero* measure. This behaviour relates to the ideas expressed in Theorem 2.3.3.   □

3.3.3 Legendre Transform

We now show that the relationship between $\widetilde{f}(y)$ and $\widetilde{\theta}(q)$ takes the form of a Legendre transform. Starting with $\widetilde{f}(y)$ as in Equation 3.6, we maximise

$$-\sum_{\omega \in \Omega_0} \phi_\omega \log_b \phi_\omega,$$

subject to the constraint that $(\phi_0, \cdots, \phi_{b-1}) \in \Psi(y)$, i.e.,

$$\sum_{\omega \in \Omega_0} \phi_\omega = 1 \qquad (3.9)$$

and

$$y = -\sum_{\omega \in \Omega_0} \phi_\omega \log_b p_\omega. \qquad (3.10)$$

Using two Lagrange multipliers, λ and q, consider the function

$$Q = -\sum_{\omega \in \Omega_0} \phi_\omega \log_b \phi_\omega + q \left(y + \sum_{\omega \in \Omega_0} \phi_\omega \log_b p_\omega \right) + \lambda \left(-1 + \sum_{\omega \in \Omega_0} \phi_\omega \right).$$

Then

$$\frac{\partial Q}{\partial \phi_\omega} = -\log_b \phi_\omega - 1 + q \log_b p_\omega + \lambda, \qquad \omega \in \Omega_0. \tag{3.11}$$

Using the two constraints and $\partial Q/\partial \phi_\omega = 0$, we solve for ϕ_ω, $\omega \in \Omega_0$. We have $s + 2$ equations, and $s + 3$ parameters; i.e., s ϕ_ω's, λ, q and y, where $s = \#\Omega_0$. We express y as a function of q, y_q. Setting Equation 3.11 equal to zero gives

$$\log_b \phi_\omega = \lambda - 1 + q \log_b p_\omega,$$

that is, $\phi_\omega = b^{\lambda-1} p_\omega^q$. From Equation 3.9 it follows that

$$\sum_{\omega \in \Omega_0} \phi_\omega = \sum_{\omega \in \Omega_0} b^{\lambda-1} p_\omega^q = 1.$$

Therefore, the maximum will occur for a given y and q when

$$\phi_\omega = \frac{p_\omega^q}{\sum_{i \in \Omega_0} p_i^q}, \qquad \omega \in \Omega_0. \tag{3.12}$$

From Equations 3.10 and 3.12

$$y_q = \frac{-1}{\left(\sum_{i \in \Omega_0} p_i^q\right)} \sum_{\omega \in \Omega_0} p_\omega^q \log_b p_\omega.$$

Then, from Equation 3.6

$$
\begin{aligned}
\widetilde{f}(y_q) &= -\sum_{\omega \in \Omega_0} \frac{p_\omega^q}{\left(\sum_{i \in \Omega_0} p_i^q\right)} \log_b \left(\frac{p_\omega^q}{\sum_{j \in \Omega_0} p_j^q}\right) \\
&= \frac{-q}{\left(\sum_{i \in \Omega_0} p_i^q\right)} \sum_{\omega \in \Omega_0} p_\omega^q \log_b p_\omega + \log_b \sum_{j \in \Omega_0} p_j^q \\
&= \frac{-q}{\left(\sum_{i \in \Omega_0} p_i^q\right)} \sum_{\omega \in \Omega_0} p_\omega^q \log_b p_\omega - \widetilde{\theta}(q). \tag{3.13}
\end{aligned}
$$

Thus,

$$\widetilde{\theta}(q) = q y_q - \widetilde{f}(y_q) = -\log_b \sum_{\omega \in \Omega_0} p_\omega^q. \tag{3.14}$$

This is a solution to the Legendre transform, given by $\widetilde{\theta}(q) = \inf_y \{qy - \widetilde{f}(y)\}$. The function y_q expresses the values of y where the infimum is attained (i.e., where the function $\widetilde{f}(y)$ is maximised) as a function of q. Note that

$$y_{\min} = \lim_{q \to \infty} y_q = \min_{\omega \in \Omega_0} (-\log_b p_\omega), \tag{3.15}$$

and

$$y_{\max} = \lim_{q \to -\infty} y_q = \max_{\omega \in \Omega_0} (-\log_b p_\omega). \tag{3.16}$$

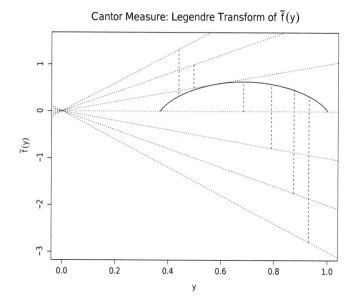

Figure 3.2 $\widetilde{f}(y)$ *for the multinomial measure with* $p_0 = \frac{1}{3}, p_1 = 0$ *and* $p_2 = \frac{2}{3}$ *(solid line).* *The slopes of the dotted lines are* $q = -3, -2, \cdots, 2, 3$. *The smallest distance between the dotted lines and the curve* $\widetilde{f}(y)$, *occurs at the value of* y_q *(see Figure 3.3), where the derivative of the function* $\widetilde{f}(y)$ *is equal to* q. *These minimum distances at* y_q *are marked by the dashed vertical lines, the lengths of which are the values of* $\widetilde{\theta}(q) = \inf_y \{qy - \widetilde{f}(y)\}$.

3.3.4 Example - Cantor Measure

Let $b = 3$ and $p_0 = \frac{1}{3}$, $p_1 = 0$ and $p_2 = \frac{2}{3}$. The Legendre transform of $\widetilde{f}(y)$ and the function y_q are plotted in Figures 3.2 and 3.3, respectively. Notice that the infimum occurs when the function $\widetilde{f}(y)$ and the lines qy have the same slope (i.e., slope $= q$). The reverse Legendre transform is plotted in Figure 3.4.   □

3.4 Fractal Dimensions

Define the mapping $g_n : [0, 1] \to \Omega^n$, where $g_n(x)$ represents the first n digits in the b-adic expansion of the number x. We want to evaluate the box or Hausdorff dimensions of the set

$$\widetilde{F}(y) = \left\{ x \in [0, 1] : \lim_{n \to \infty} Y_n(g_n(x)) = y \right\}.$$

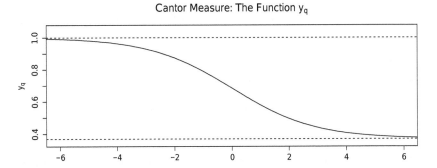

Figure 3.3 *The function y_q for the multinomial measure with $p_0 = \frac{1}{3}, p_1 = 0$ and $p_2 = \frac{2}{3}$. The slope of the function $\widetilde{f}(y)$ (see Figure 3.2) at $y = y_q$ is q. The two asymptotes, $y_{\min} = \lim_{q \to \infty} y_q$ and $y_{\max} = \lim_{q \to -\infty} y_q$, are the values at which the function $\widetilde{f}(y)$ cuts the horizontal axis.*

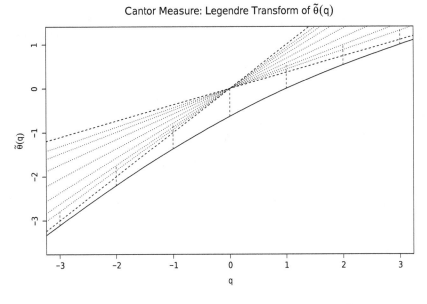

Figure 3.4 *$\widetilde{\theta}(q)$ is plotted (solid line). As q becomes large and positive, $\widetilde{\theta}(q)$ tends to the line $qy_{\min}$, and as q becomes large and negative, $\widetilde{\theta}(q)$ tends to the line $qy_{\max}$ (diagonal dashed lines). The dotted lines each have slope $y = y_q, q = -3, -2, \cdots, 2, 3$ (see Figure 3.3). The derivative of $\widetilde{\theta}(q)$ is y_q. The smallest distance (vertical dashed lines), $\widetilde{f}(y)$, between the curve $\widetilde{\theta}(q)$ and a line qy (i.e., a line with slope y) will occur at q when the slope of the line is y_q; i.e., $\widetilde{f}(y_q) = qy_q - \widetilde{\theta}(q)$ is the solution to $\widetilde{f}(y) = \inf_q\{qy - \widetilde{\theta}(q)\}$.*

3.4.1 Eggleston's Theorem (Eggleston, 1949)

Let $\Phi_\alpha(x)$ be the asymptotic proportion of digits that equal α in the base b expansion of x, and

$$M(\phi_0, \cdots, \phi_{b-1}) = \{x \in [0,1] : \Phi_0(x) = \phi_0, \cdots, \Phi_{b-1}(x) = \phi_{b-1}\},$$

where $\sum_{\omega \in \Omega} \phi_\omega = 1$. Then the Hausdorff dimension of $M(\phi_0, \cdots, \phi_{b-1})$ is

$$\dim_H M(\phi_0, \cdots, \phi_{b-1}) = - \sum_{\omega \in \Omega_0} \phi_\omega \log_b \phi_\omega. \tag{3.17}$$

See also Falconer (1990, §10.1).   □

It is known that almost all numbers (in the Lebesgue sense) are *normal* to all bases; i.e., the set $M(\frac{1}{b}, \cdots, \frac{1}{b})$ has Lebesgue measure one and, therefore, dimension one. This is consistent with Equation 3.17. Note that if $\phi_\omega > 0$ for all $\omega \in \Omega$, the set $M(\phi_0, \cdots, \phi_{b-1})$ is dense on $[0,1]$ and, hence, $\dim_B M(\phi_0, \cdots, \phi_{b-1})$ is one. This is an unfortunate deficiency of the box counting dimension (see Proposition A.3.4 and Example A.3.5).

3.4.2 Example

Consider the special case where $s = \#\Omega_0 = 2$, then $\exists i, j : i \neq j$ with $p_i = 1 - p_j$, and such that all $p_m = 0$ where $m \neq i$ and $m \neq j$. Then it follows from Equation 3.2 that

$$Y_n(\omega) = \begin{cases} -\Phi_i(\omega) \log_b \left(\dfrac{p_i}{p_j}\right) - \log_b p_j & \text{if } p_i \neq p_j \\ -\log_b p_j & \text{if } p_i = p_j. \end{cases}$$

In this case it can be seen that, given p_i and p_j and $p_i \neq p_j$, $\widetilde{F}(y)$ can be expressed in terms of unique values of $\Phi_i(\omega)$ and hence $\Phi_j(\omega)$. That is,

$$\widetilde{F}(y) = \{x \in [0,1] : \Phi_i(x) = \phi_i, \Phi_j(x) = \phi_j, \text{ and } \Phi_m(x) = 0 \ \forall m \neq i, j\}.$$

Note that y is related to ϕ_i and ϕ_j through Equation 3.10, i.e., $y = -\phi_i \log_b p_i - \phi_j \log_b p_j$. The set $\Psi(y)$, given by Equation 3.3, consists of a *unique point* if $s = 2$ and $p_i \neq p_j$ and, hence, it follows from Equation 3.6 that

$$\widetilde{f}(y) = - \sum_{\omega \in \Omega_0} \phi_\omega \log_b \phi_\omega = \begin{cases} -\phi_i \log_b \phi_i - \phi_j \log_b \phi_j & p_i \neq p_j \\ -\log_b p_j & p_i = p_j. \end{cases}$$

From Equation 3.17, $\dim_H \widetilde{F}(y) = \widetilde{f}(y)$.   □

3.5 Point Centred Construction

The method of describing the multifractal characteristics of the multinomial measure in this chapter is based on a system of lattice (fixed width) coverings. What is the relationship between these results and those that one may derive from a system of point centred coverings? We briefly investigate this question in this section.

3.5.1 Example

Consider the Cantor measure, as in Example 1.2.1, with $b = 3$, $p_0 = 1 - p_2$ and $p_1 = 0$. Here we let p_0 take any value between zero and one. The measure is supported by the Cantor set, which has the characteristic middle third gaps. At each division, the subintervals on the left and right receive p_0 and p_2 of the mass respectively. Let μ_n be the measure distribution on $\mathcal{K}_n$ at the nth stage of this process (see Equation 3.1).

The expectation $E\left[\mu^{q-1}[S_\delta(X)]\right]$, where X is a random variable selected from the distribution μ, can be calculated by initially evaluating $E\left[\mu_n^{q-1}[S_\delta(X)]\right]$. Each subinterval with non-zero measure has a uniform density (as in Example 2.4.4), and so one can express the integral, which we will denote as $A_n(q, \delta)$, as a sum over these intervals, i.e.,

$$A_n(q, \delta) = \sum_{\omega \in \Omega_0^n} \int_{x \in J_n(\omega)} \mu_n^{q-1}[S_\delta(x)]\mu_n(dx),$$

for sufficiently small δ (i.e., $\delta < 3^{-n}$); where Ω_0^n contains all permutations of length n of the digits 0 and 2, and $J_n(\omega)$ is the subinterval based on the b-adic sequence $\omega \in \Omega_0^n$. However,

$$D_q \neq \lim_{\delta \to 0} \frac{\log\left\{\lim_{n \to \infty} A_n(q, \delta)\right\}}{\log \delta}$$

because when the first limit in n is taken, the gaps are getting smaller. If δ remains fixed, then there reaches a point where the covering spheres have radius δ that completely spans the gap, and covers part of the support on the other side. The limits cannot be interchanged either. For a fixed finite value of n, μ_n consists of a collection of essentially uniform type measures separated by gaps. Taking the limit in δ on this measure will simply give $D_q = 1$ for all q, as in Example 2.4.4.

Consider just the case where $q = 2$. In order to find an analytic solution for D_2, we need to return to the ideas at the start of §2.4, in particular, Equation 2.4. Let the independent random variables X_1 and X_2 be drawn from the distribution (μ_n) on $\mathcal{K}_n$. Then from Equation 2.4,

$$\int \mu_n[S_\delta(x)]\mu_n(dx) = \Pr\{|X_1 - X_2| \leq \delta\}.$$

Let $h_n(y)$ be the density of $X_1 - X_2$ and $g_n(y)$ be the density of $|X_1 - X_2|$. At the start, the distribution on $\mathcal{K}_0$ is just the uniform distribution, and so

$$\begin{aligned} h_0(y) &= 1 - |y| & -1 \leq y \leq 1 \\ g_0(y) &= 2 - 2y & 0 \leq y \leq 1. \end{aligned}$$

For each n, $h_n(y)$ is a scaled copy of $h_{n-1}(y)$, in that $h_{n-1}(y)$ is contracted onto the interval $\left[-\frac{1}{3}, \frac{1}{3}\right]$. Copies of this contraction are also placed on the intervals $\left[-1, -\frac{1}{3}\right]$ and $\left[\frac{1}{3}, 1\right]$. The inner copy contains $p_0^2 + p_2^2$ of the mass, the outer two

$p_0 p_2$ each, thus $h_n(y)$ satisfies the recurrence relation

$$h_n(y) = \begin{cases} 3(p_0^2 + p_2^2)h_{n-1}(3y) & 0 \le |y| \le \frac{1}{3} \\ 3p_0 p_2 h_{n-1}(3|y| - 2) & \frac{1}{3} \le |y| \le 1. \end{cases}$$

Consequently,

$$g_n(y) = \begin{cases} 3(p_0^2 + p_2^2)g_{n-1}(3y) & 0 \le y \le \frac{1}{3} \\ 3p_0 p_2 g_{n-1}(2 - 3y) & \frac{1}{3} \le y \le \frac{2}{3} \\ 3p_0 p_2 g_{n-1}(3y - 2) & \frac{2}{3} \le y \le 1. \end{cases}$$

The probability distribution of the absolute interpoint distances then satisfies the following recurrence relation:

$$G_n(y) = \begin{cases} (p_0^2 + p_2^2)G_{n-1}(3y) & 0 \le y \le \frac{1}{3} \\ 1 - p_0 p_2 - p_0 p_2 G_{n-1}(2 - 3y) & \frac{1}{3} \le y \le \frac{2}{3} \qquad (3.18) \\ 1 - p_0 p_2 + p_0 p_2 G_{n-1}(3y - 2) & \frac{2}{3} \le y \le 1. \end{cases}$$

It follows that $G_n(y) = (p_0^2 + p_2^2)^n G_0(3^n y)$ for $0 < y < 3^{-n}$, where $G_0(y) = 2y - y^2$. Let $G_\infty(y)$ be the limiting probability distribution, then

$$D_2 = \lim_{y \to 0} \frac{\log \Pr\{|X_1 - X_2| \le y\}}{\log y} = \lim_{y \to 0} \frac{G_\infty(y)}{\log y}.$$

By construction, $G_\infty(3^{-n}) = (p_0^2 + p_2^2)^n G_0(1) = (p_0^2 + p_2^2)^n$. Therefore, for all y such that $3^{-(n+1)} \le y \le 3^{-n}$,

$$(p_0^2 + p_2^2)^{n+1} \le G_\infty(y) \le (p_0^2 + p_2^2)^n,$$

consequently

$$\frac{\log[(p_0^2 + p_2^2)^{n+1}]}{\log(3^{-n})} \ge \frac{\log G_\infty(y)}{\log y} \ge \frac{\log[(p_0^2 + p_2^2)^n]}{\log(3^{-(n+1)})}.$$

Taking limits in n implies that

$$D_2 = \lim_{y \to 0} \frac{\log G_\infty(y)}{\log y} = -\log_3(p_0^2 + p_2^2) = \tilde{D}_2.$$

$\square$

In fact, it is easy to see that the above argument holds for any multinomial measure with gaps for all interpoint distances of order q, i.e., $D_q = \tilde{D}_q$ for $q = 2, 3, \cdots$.

PART II

MULTIFRACTAL FORMALISM USING LARGE DEVIATIONS

CHAPTER 4

Lattice Based Multifractals

4.1 Introduction

In this chapter, sufficient conditions are deduced under which the multifractal lattice formalism holds, in both weak and strong senses (Definition 2.5.4). In this section an outline of the overall argument is given.

Let $\mathcal{X}$ be a subset of $\mathbb{R}^d$ and $\mathcal{B}(\mathcal{X})$ be the Borel subsets of $\mathcal{X}$. We are interested in the *probability* space $(\mathcal{X}, \mathcal{B}(\mathcal{X}), \mu)$, where the measure μ is assumed to be non-atomic and may be concentrated on a subspace of $\mathcal{X}$ of lower dimension than d. In this chapter we consider the case where $\mathcal{X}$ is covered by a succession of lattices of d-dimensional boxes that are half open to the right and of width δ_n. The kth box is denoted by $B_{\delta_n}(k)$, where $k \in K_n$ and $K_n = \{k : \mu[B_{\delta_n}(k)] > 0\}$. We evaluate successive lattice coverings for some sequence $\{\delta_n\}$, where $\delta_n \to 0$ as $n \to \infty$. Using this system of lattices, we evaluate the local behaviour of the measure μ denoted by Y_n which, as in Equation 2.10, will be the ratio of the logarithm of the measure within a given box divided by the logarithm of the box width.

Note that in the multinomial example of Chapter 3, the covering consisted of closed intervals, i.e., closed versions of the b-adic intervals. The reason for this discrepancy will be discussed in Example 4.5.11.

To appeal to the theory of large deviations, it is easiest to define an auxiliary probability space $(\mathcal{X}, \mathcal{F}_n, \nu^{(n)})$, where $\mathcal{F}_n$ is the sub σ-field generated by the lattice boxes at the nth stage, and $\nu^{(n)}$ is an arbitrary probability measure describing the way in which the boxes are sampled. We then treat Y_n as a random variable, where $Y_n : \mathcal{X} \to \mathbb{R}$.

The weak multifractal formalism involving the Legendre transform relationships can be deduced by appealing to the Gärtner-Ellis theorem of large deviations. In this case, Y_n converges to a degenerate point that is dependent on the particular form of the sampling measure $\nu^{(n)}$. The convergence is similar to a weak law of large numbers, but at an exponential rate. It will be shown that the multifractal spectrum $\widetilde{f}(y)$ is related to the entropy function which describes rates of convergence, and that $\widetilde{\theta}(q)$ is related to a rescaled cumulant generating function of Y_n. The required mathematical framework is established in §4.2 for an arbitrary sampling measure. In §4.3 the case of the uniform sampling measure is discussed, and in §4.4 a family of sampling measures is defined in such a way that the rescaled cumulant generating functions and entropy functions for individual members of the family are related via simple shift type transformations.

In §4.5 we determine sufficient conditions under which $\widetilde{f}(y)$ is equal to the Hausdorff dimension of the local behaviour of μ. In order to achieve this, we require the sampling measure $\nu^{(n)}$ to have an extension to $\nu^{(\infty)}$, where $\nu^{(\infty)}$ is the inductive limit of $\nu^{(n)}$. That is, we now are considering the behaviour of Y_n on the probability space $(\mathcal{X}, \mathcal{B}(\mathcal{X}), \nu^{(\infty)})$. In this situation, Y_n convergences to the degenerate point almost surely. It also turns out to be easier to characterise the local behaviour of the measure $\nu^{(\infty)}$ rather than that of μ. Since we have almost sure convergence, then the sets describing the local behaviour of μ can be shown to be the same as certain sets describing the local behaviour of $\nu^{(\infty)}$, apart from a set of measure zero. Thus, the problem essentially reduces to determining the Hausdorff dimension of the local behaviour of $\nu^{(\infty)}$. This can be deduced by appealing to a theorem by Billingsley (1965), and then relating back to the local behaviour of the measure μ.

4.2 Large Deviation Formalism

In the multinomial example of Chapter 3, it was seen that one way of characterising the measure μ was to describe its local behaviour, Y_n, given by Equation 3.2. However, treating Y_n as a random variable, it was shown that it had a degenerate distribution. Information could still be gleaned by characterising the convergence rate of Y_n. This is the problem described by the theory of large deviations. An introduction to the theory of large deviations is given in Appendix B where statements of results that are used in this chapter can be found. In this section, a general probability framework is defined for Y_n, which will form the basic structure for discussions in the subsequent sections. We start with the auxiliary probability space $(\mathcal{X}, \mathcal{F}_n, \nu^{(n)})$, where $\nu^{(n)}$ is some arbitrary *sampling measure*.

In order to describe the behaviour of μ, we define a sequence of measurable mappings $\{U_n\}, n = 1, 2, \cdots$, where

$$U_n : (\mathcal{X}, \mathcal{F}_n) \to (\mathbb{R}, \mathcal{B}(\mathbb{R})), \qquad (4.1)$$

as

$$U_n(x) = \begin{cases} -\log \mu[B_{\delta_n}(k(x))] & \text{if } k(x) \in K_n \\ 0 & \text{if } k(x) \notin K_n, \end{cases}$$

where $k(x)$ is interpreted as the index k of the box that contains the point x. Also define the functions $\{Y_n\}, n = 1, 2, \cdots$ as

$$Y_n(x) = \frac{U_n(x)}{-\log \delta_n}.$$

Note that Y_n has been defined in terms of $x \in \mathcal{X}$, but is the same for all x in the same lattice box. We want to describe the behaviour of the measure μ by investigating the size of the set

$$\{B_{\delta_n} : -\epsilon < Y_n - y \le \epsilon\} = \bigcup_{k \in K_n(y, \epsilon)} B_{\delta_n}(k),$$

see Definition 2.5.1, the size being the value of the measure $\nu^{(n)}$ on that set; i.e., $\nu^{(n)}\{B_{\delta_n} : -\epsilon < Y_n - y \le \epsilon\}$ for various values of y. Different sampling measures will presumably select different characteristics.

Using the theory of large deviations to describe the mapping Y_n involves the following main steps. One first defines the cumulant generating function, denoted by $C(q)$. This is essentially a more general version of $\widetilde{\theta}(q)$. The theory of large deviations is based on arguments of convexity. Therefore, to proceed further we need to assume that $C(q)$ has the required properties that will be specified in the Extended Hypotheses. Then one defines the *entropy* function $I(y)$ which is simply the Legendre transform of $C(q)$. Again, $I(y)$ is a more general version of $\widetilde{f}(y)$. If sufficient conditions hold, the degenerate point to which Y_n converges can be easily calculated using $C(y)$. In fact, the convergence occurs at an exponential rate. The entropy function $I(y)$ describes the rates of convergence, which will be characterised by the large deviation bounds.

4.2.1 Rescaled Cumulant Generating Function

From Equation B.5, the *Rescaled Cumulant Generating Function* is $C(q)$, if the limit $(n \to \infty)$ exists, where

$$C(q) = \lim_{n \to \infty} \frac{1}{-\log \delta_n} \log \mathrm{E}_n[e^{qU_n}],$$

and E_n indicates that the expectation is to be taken with respect to the sampling measure $\nu^{(n)}$. Therefore,

$$C(q) = \lim_{n \to \infty} \frac{1}{-\log \delta_n} \log \left(\sum_{k \in K_n} \mu^{-q}[B_{\delta_n}(k)]\nu^{(n)}[B_{\delta_n}(k)] \right), \qquad (4.2)$$

for $q \in \mathbb{R}$. Note that we allow $+\infty$ as a limit value. Further, let

$$I(y) = \sup_{q \in \mathbb{R}}\{qy - C(q)\}, \qquad y > 0. \qquad (4.3)$$

$I(y)$ may take the value of $+\infty$, and is referred to as the Legendre transform of $C(q)$.   □

It can be seen that $C(q)$ is similar in nature to the global average $\widetilde{\theta}(q)$ defined in Definition 2.3.1. In fact $\widetilde{\theta}(q)$ is the special case when the sampling measure gives each box equal weight. We consider this case in §4.3.

4.2.2 Extended Hypotheses

Assume that there exists a sequence $\{\delta_n\}$, with $\delta_n \to 0$ as $n \to \infty$, such that

1. $C(q)$ satisfies Hypotheses B.3.8, and

2. $C(q)$ is differentiable on the interior of its domain, $\mathrm{int}\mathcal{D}(C) = \mathrm{int}\{q \in \mathbb{R} : C(q) < \infty\}$, and $C(q)$ is steep (Definition B.2.1).   □

The arguments of this chapter require that $C(q)$ satisfies the above Extended Hypotheses. Generally, $C(q)$ will not satisfy these conditions for any arbitrary sequence $\{\delta_n\}$. In the case of the multinomial measures in Chapter 3, $\delta_n = b^{-n}$.

Given that $C(q)$ is sufficiently well behaved, as in Extended Hypotheses 4.2.2 above, then it can be shown that Y_n converges exponentially to a point y_0 that can be deduced from the rescaled cumulant generating function. Say Y_n represents a partial mean, i.e., a partial sum divided by n. Then exponential convergence (Definition B.3.13) is just a weak law of large numbers, but where the convergence occurs at an exponential rate. A more familiar situation where these ideas are applied to sums of independent normal random variables can be found in Appendix B, Example B.2.4. This similarity with the weak law of large numbers will be described further in §4.3.

4.2.3 Theorem - Exponential Convergence

If $C(q)$ satisfies the Extended Hypotheses 4.2.2, then $I(y)$ has a unique minimum at $y_0 = C'(0)$ such that $I(y_0) = 0$. Further, $Y_n \xrightarrow{\text{exp}} y_0$ with respect to $\nu^{(n)}$ (see Definition B.3.13), and

$$C(q) = \sup_y \{qy - I(y)\}. \tag{4.4}$$

Proof. Given Hypotheses B.3.8, it follows from Theorem B.3.10 that $\inf_y I(y) = 0$. Further, since $C(q)$ is differentiable (Extended Hypotheses 4.2.2) on $\text{int}\mathcal{D}(C)$ which is non-empty and contains the point $q = 0$ (Hypotheses B.3.8), it follows from Theorem B.3.14 that $I(y)$ has a unique minimum at $y_0 = C'(0)$ and $Y_n \xrightarrow{\text{exp}} y_0$. Equation 4.4 follows from Theorem B.3.17. □

Since $I(y) > 0$ for all $y \neq y_0$, $I(y)$ is the powerlaw exponent (i.e., $\delta_n^{I(y)}$) with which $\nu^{(n)}\{B_{\delta_n} : -\epsilon < Y_n - y \leq \epsilon\}$ approaches zero for all $y \neq y_0$. We express this formally in the next theorem.

4.2.4 Theorem - Large Deviation Bounds

If $C(q)$ satisfies the Extended Hypotheses 4.2.2, then

$$\lim_{n \to \infty} \frac{\log \nu^{(n)}\{B_{\delta_n} : -\epsilon < Y_n - y \leq \epsilon\}}{-\log \delta_n} = - \inf_{z \in (y-\epsilon, y+\epsilon]} I(z)$$

where $\epsilon > 0$ and $(y - \epsilon, y + \epsilon] \subseteq \mathcal{D}(I)$. Further, taking limits in ϵ gives

$$\lim_{\epsilon \to 0} \lim_{n \to \infty} \frac{\log \nu^{(n)}\{B_{\delta_n} : -\epsilon < Y_n - y \leq \epsilon\}}{-\log \delta_n} = -I(y) \quad \text{for } y \in \text{int}\mathcal{D}(I). \tag{4.5}$$

Proof. Given that (1) of the Extended Hypotheses 4.2.2 is valid, then the Gärtner-Ellis Theorem B.3.10 ensures that the upper large deviation bound is valid. Similarly, given (2) of the Extended Hypotheses 4.2.2, the Gärtner-Ellis Theorem

B.3.10 ensures that the lower large deviation bound is valid, and that $I(y)$ is convex and closed. It then follows from Lemma B.3.12 that $I(y)$ is continuous on $\mathcal{D}(I)$. The first equation then follows from Theorem B.3.6. Equation 4.5 follows by taking limits in ϵ. $\qquad\square$

4.3 Uniform Spatial Sampling Measure

In this section, the results of §4.2 are applied to the case where the sampling measure is *uniform*. We mean uniform in the sense that, in the pre-limiting case (finite n in $\nu^{(n)}$), each box that contains non-zero μ mass is given the same weight by the sampling measure. For this case denote the sampling measure $\nu^{(n)}$ as $\nu_0^{(n)}$ where

$$\nu_0^{(n)}[B_{\delta_n}(k)] = \begin{cases} \dfrac{1}{\#K_n} & k \in K_n \\[2mm] 0 & k \notin K_n. \end{cases}$$

Similarly, denote $C(q)$ and $I(y)$ as $C_0(q)$ and $I_0(y)$, respectively. This is the situation most often considered in the literature.

4.3.1 Theorem - Rescaled Cumulant Generating Function

If $\widetilde{\theta}(0)$ exists, or equivalently $-1 \in \mathcal{D}(C_0)$, then

$$C_0(q) = \widetilde{\theta}(0) - \widetilde{\theta}(-q), \qquad (4.6)$$

where $\widetilde{\theta}(q)$ is given in Definition 2.3.1.

Proof. From Equation 4.2,

$$C_0(q) = \lim_{n \to \infty} \frac{1}{-\log \delta_n} \log \left(\frac{1}{\#K_n} \sum_{k \in K_n} \mu^{-q}[B_{\delta_n}(k)] \right).$$

Since $C_0(-1) = \widetilde{\theta}(0)$, then $-1 \in \mathcal{D}(C_0)$ iff $\widetilde{\theta}(0)$ exists and is finite. Therefore,

$$\begin{aligned} C_0(q) &= \lim_{n \to \infty} \frac{\log \#K_n}{\log \delta_n} - \lim_{n \to \infty} \frac{1}{\log \delta_n} \log \sum_{k \in K_n} \mu^{-q}[B_{\delta_n}(k)] \\ &= \widetilde{\theta}(0) - \widetilde{\theta}(-q) \quad \text{if} - 1 \in \mathcal{D}(C_0) \end{aligned}$$

for $q \in \mathcal{D}(C_0)$. $\qquad\square$

4.3.2 Corollary - Exponential Convergence

If $C_0(q)$ satisfies the Extended Hypotheses 4.2.2 then $Y_n \xrightarrow{\text{exp}} y_0 = C_0'(0)$. Further, $I_0(y_0) = 0$ is the unique minimum of $I_0(y)$.

Proof. Since $C_0(q)$ satisfies the Extended Hypotheses 4.2.2, $0 \in \text{int} \mathcal{D}(C_0)$. Further, C_0 is differentiable on the interior of $\mathcal{D}(C_0)$, hence y_0 exits. The result then follows from Theorem 4.2.3.   □

4.3.3 Example

Consider again the multinomial example of Chapter 3. In this case, the boxes or subintervals at the nth stage are more easily characterised by ω, a string of digits of length n, i.e., $\omega \in \Omega^n$. $J_n(\omega)$ is the interval characterised by ω. Intervals with non-zero measure are characterised by $\omega \in \Omega_0^n$ and have width $\delta_n = b^{-n}$. Therefore, for $\omega = (\omega_1, \cdots, \omega_n) \in \Omega_0^n$ where $\omega_i \in \Omega_0$,

$$Y_n(\omega) = \frac{\log \mu[J_n(\omega)]}{\log \delta_n} = \frac{\log \left(\prod_{i=1}^{n} p_{\omega_i} \right)}{\log b^{-n}} = \frac{1}{n} \sum_{i=1}^{n} \log_b p_{\omega_i},$$

and hence, in this example, Y_n can be written explicitly as a partial sum. This converges exponentially to y_0 where, from Equation 4.6,

$$y_0 = C_0'(0) = \frac{-1}{\#\Omega_0} \sum_{\omega \in \Omega_0} \log_b p_\omega.$$

Hence if we sample with equal weight, then $Y_n \xrightarrow{\exp} y_0$, thus satisfying a weak law of large numbers.   □

4.3.4 Corollary - Large Deviation Bounds

Given that $C_0(q)$ satisfies the Extended Hypotheses 4.2.2, then

$$\lim_{\epsilon \to 0} \lim_{n \to \infty} \frac{\log \nu_0^{(n)} \{ B_{\delta_n} : -\epsilon < Y_n - y \leq \epsilon \}}{-\log \delta_n} = -I_0(y) \qquad \text{for } y \in \text{int} \mathcal{D}(I_0).$$

Further, if $-1 \in \mathcal{D}(C_0)$, then

$$I_0(y) = -\widetilde{\theta}(0) - \widetilde{f}(y), \tag{4.7}$$

where $\widetilde{f}(y)$ is given by Equation 2.7.

Proof. The double limit follows from Theorem 4.2.4. Equation 4.7 follows from Equation 4.3 and the relationship

$$\nu_0^{(n)} \{ B_{\delta_n} : -\epsilon < Y_n - y \leq \epsilon \} = \frac{\#K_n(y, \epsilon)}{\#K_n}, \tag{4.8}$$

where $K_n(y, \epsilon)$ is given in Definition 2.5.4.   □

The following corollary establishes conditions under which μ is a multifractal measure in a weak sense (Definition 2.5.4).

4.3.5 Corollary - Legendre Transform Pair

Given that $C_0(q)$, or equivalently $\widetilde{\theta}(0) - \widetilde{\theta}(-q)$, satisfies the Extended Hypotheses 4.2.2 and $-1 \in \mathcal{D}(C_0)$, then

$$\widetilde{f}(y) = \inf_q \left\{ qy - \widetilde{\theta}(q) \right\}, \tag{4.9}$$

and

$$\widetilde{\theta}(q) = \inf_y \left\{ qy - \widetilde{f}(y) \right\}. \tag{4.10}$$

Proof. Note that $I_0(y)$ is *defined* as the Legendre transform of $C_0(q)$, i.e., $I_0(y) = \sup_q \{ qy - C_0(q) \}$. Given that $C_0(q)$ satisfies the Extended Hypotheses 4.2.2 and $-1 \in \mathcal{D}(C_0)$, then it follows from Equations 4.6 and 4.7 that

$$-\widetilde{\theta}(0) - \widetilde{f}(y) = \sup_q \{ qy + -\widetilde{\theta}(0) + \widetilde{\theta}(-q) \},$$

and hence Equation 4.9. Similarly from Equation 4.4, $C_0(q) = \sup_y \{ qy - I_0(y) \}$ and using the same substitutions gives Equation 4.10. $\qquad\square$

Heuristically, one might argue that if $\delta_n^{-\widetilde{f}(y)}$ is roughly the number of boxes where μ is of the order δ_n^y, then

$$\sum_{k \in K_n} \mu^q[B_{\delta_n}(k)] \sim \sum_y \delta_n^{qy} \delta_n^{-\widetilde{f}(y)} \sim \delta_n^{\inf_y \{ qy - \widetilde{f}(y) \}}$$

and therefore

$$\widetilde{\theta}(q) = \inf_y \left\{ qy - \widetilde{f}(y) \right\}.$$

These relationships are not so simple, and depend on arguments of convexity. Consider an example where $\widetilde{\theta}(q)$ is not differentiable, and the Legendre transform relationships given in this chapter do not hold.

4.3.6 Example (Holley & Waymire, 1992)

Let $\mathcal{X} = [0,1] \times [0,1]$. Initially, cover $\mathcal{X}$ with a $b \times b$ δ_n-mesh. At each iteration, partition each box into b^2 further boxes. At the nth iteration $\delta_n = b^{-n}$ and $\#K_n = b^{2n}$. Let $\mu = \frac{1}{2}\lambda_1 \times \delta_0 + \frac{1}{2}\lambda_2$, where λ_1 is a one-dimensional Lebesgue measure on $[0,1]$, δ_0 is the Dirac unit mass at zero and λ_2 is the Lebesgue measure on $[0,1] \times [0,1]$. Then

$$
\begin{aligned}
\sum_{k \in K_n} \mu^q[B_{\delta_n}(k)] &= b^n \left(\frac{1}{2}b^{-n} + \frac{1}{2}b^{-2n} \right)^q + (b^{2n} - b^n) \left(\frac{1}{2}b^{-2n} \right)^q \\
&= \left(\frac{1}{2}b^{-2n} \right)^q b^n (b^n + 1)^q + b^n (b^n - 1) \left(\frac{1}{2}b^{-2n} \right)^q.
\end{aligned}
$$

It is easy to see that if $q < 1$, the summation tends to infinity as $n \to \infty$. The purpose of the exercise is to determine the *rate* at which this occurs.

There are two situations, $q \geq 1$ and $q \leq 1$. Firstly consider $q \leq 1$, so

$$\sum_{k \in K_n} \mu^q[B_{\delta_n}(k)] = \left(\frac{1}{2}b^{-2n}\right)^q b^{2n}\left[b^{n(q-1)}(1 + b^{-n})^q + (1 - b^{-n})\right],$$

$$(4.11)$$

and thus the term in square brackets tends to one as $n \to \infty$. Therefore,

$$\begin{aligned}
\tilde{\theta}(q) &= \lim_{n \to \infty} \frac{\log \sum_{k \in K_n} \mu^q[B_{\delta_n}(k)]}{\log(b^{-n})} \\
&= 2(q-1) - \lim_{n \to \infty} \frac{1}{n} \log_b \left[b^{n(q-1)}(1 + b^{-n})^q + (1 - b^{-n})\right] \\
&= 2(q-1) \quad \text{for } q \leq 1.
\end{aligned}$$

For $q \geq 1$, we rearrange Equation 4.11 to give

$$\sum_{k \in K_n} \mu^q[B_{\delta_n}(k)] = \left(\frac{1}{2}b^{-2n}\right)^q b^{2n}b^{n(q-1)}\left[(1 + b^{-n})^q + (1 - b^{-n})b^{-n(q-1)}\right].$$

Similarly,

$$\begin{aligned}
\tilde{\theta}(q) &= \lim_{n \to \infty} \frac{\log \sum_{k \in K_n} \mu^q[B_{\delta_n}(k)]}{\log(b^{-n})} \\
&= (q-1) + \lim_{n \to \infty} \frac{1}{n} \log_b \left[(1 + b^{-n})^q + (1 - b^{-n})b^{-n(q-1)}\right] \\
&= q - 1 \quad \text{for } q \geq 1.
\end{aligned}$$

We thus have

$$\tilde{\theta}(q) = \begin{cases} 2(q-1) & \text{if } q \leq 1 \\ q - 1 & \text{if } q \geq 1. \end{cases}$$

It can be seen from Equation 2.7 that

$$\tilde{f}(y) = \begin{cases} 1 & \text{if } y = 1 \\ 2 & \text{if } y = 2 \\ -\infty & \text{otherwise,} \end{cases}$$

and

$$\dim_H \left\{x : \lim_{n \to \infty} \frac{\log \mu[B_{\delta_n}(x)]}{\log \delta_n} = y\right\} = \begin{cases} 1 & \text{if } y = 1 \\ 2 & \text{if } y = 2 \\ 0 & \text{otherwise.} \end{cases}$$

Note that $\tilde{\theta}(q)$ is continuous but *not differentiable* and that

$$\tilde{f}(y) \neq \inf_q \left\{qy - \tilde{\theta}(q)\right\} = \begin{cases} y & \text{if } 1 \leq y \leq 2 \\ -\infty & \text{otherwise.} \end{cases}$$

□

4.4 A Family of Sampling Measures

The points to which the sequence Y_n converges exponentially are dependent on the sampling measure. By considering a family of sampling measures, we get a family of limit points. The relationships described in this section are not required to determine conditions for which μ is a multifractal measure in a weak sense (see Corollary 4.3.5). However, they do form the basis of the ideas that are required to describe requirements for μ to be a multifractal measure in a strong sense.

4.4.1 Definition

Define the set function M_m as

$$M_m(B) = \begin{cases} \mu^m(B) & \text{if } \mu(B) \neq 0 \\ 0 & \text{if } \mu(B) = 0, \end{cases} \tag{4.12}$$

where $m \in \mathbb{R}$ and $B \in \mathcal{B}(\mathcal{X})$. Then the family of sampling measures is defined as

$$\nu_m^{(n)}(A) = \frac{\sum_{j \in K_n} M_m[A \cap B_{\delta_n}(j)]}{\sum_{k \in K_n} \mu^m[B_{\delta_n}(k)]} \qquad A \in \mathcal{F}_n, \tag{4.13}$$

where m is a number such that $-m \in \text{int}\mathcal{D}(C_0)$.   □

In this case, denote $C(q)$ and $I(y)$ as $C_m(q)$ and $I_m(y)$, respectively. The case where $m = 0$ is the uniform measure considered in the previous section. When $m = 1$, each box receives weight proportional to the probability of the process visiting that box. When $m < 0$, the boxes are sampled with an inverse probability to the amount of time the process spends in each box.

4.4.2 Theorem - Rescaled Cumulant Generating Function

If $-m \in \mathcal{D}(C_0)$, then

$$C_m(q) = C_0(q - m) - C_0(-m), \tag{4.14}$$

hence $\mathcal{D}(C_m) = \{q : C_0(q - m) < \infty\}$. Also,

$$I_m(y) = C_0(-m) + my + I_0(y). \tag{4.15}$$

Proof. Equation 4.14 follows by inserting $\nu_m^{(n)}$ into Equation 4.2 and simplifying. From Equations 4.3 and 4.14,

$$\begin{aligned} I_m(y) &= \sup_q \{qy - C_0(q - m) + C_0(-m)\} \\ &= C_0(-m) + \sup_q \{(q + m)y - C_0(q)\} \\ &= C_0(-m) + my + I_0(y). \end{aligned}$$

□

From Equations 4.14 and 4.15, it can be seen that $C_m(q)$ and $I_m(y)$ are related to $C_0(q)$ and $I_0(y)$, respectively, by a shift type transformation. Using Equations 4.7 and 4.6 we can further relate $C_m(q)$ and $I_m(y)$ to $\widetilde{\theta}(q)$ and the multifractal spectrum $\widetilde{f}(y)$, respectively.

4.4.3 Lemma

Given that $C_0(q)$ satisfies the Extended Hypotheses 4.2.2 and $-m \in \operatorname{int}\mathcal{D}(C_0)$, then $C_m(q)$ satisfies the Extended Hypotheses 4.2.2. Further, $I_m(y)$ is continuous on $\mathcal{D}(I_0)$.

Proof. Equations 4.14 and 4.15 hold if $-m \in \mathcal{D}(C_0)$. However, the hypotheses require that $0 \in \operatorname{int}\mathcal{D}(C_m)$, which is satisfied if $-m \in \operatorname{int}\mathcal{D}(C_0)$. The other requirements on $C_m(q)$ follow directly from Equation 4.14. Continuity of $I_m(y)$ follows from Equation 4.15 and continuity of I_0.   □

4.4.4 Corollary - Exponential Convergence

If $C_0(q)$ satisfies the Extended Hypotheses 4.2.2, and $-m \in \operatorname{int}\mathcal{D}(C_0)$, then $Y_n \xrightarrow{\exp} y_m$ with respect to $\nu_m^{(n)}$, where $y_m = C'_m(0) = C'_0(-m)$. Further, $I_m(y_m) = 0$ is the unique minimum of $I_m(y)$.

Proof. Follows directly Theorem 4.2.3 and Lemma 4.4.3.   □

4.4.5 Corollary - Large Deviation Bounds

Given that $C_0(q)$ satisfies the Extended Hypotheses 4.2.2 and $-m \in \operatorname{int}\mathcal{D}(C_0)$, then

$$\lim_{\epsilon \to 0} \lim_{n \to \infty} \frac{\log \nu_m^{(n)}\{B_{\delta_n} : -\epsilon < Y_n - y \le \epsilon\}}{-\log \delta_n} = -I_m(y) \quad \text{for } y \in \operatorname{int}\mathcal{D}(I_0).$$

Further, if $-1 \in \mathcal{D}(C_0)$ then

$$I_m(y) = -\widetilde{\theta}(m) + my - \widetilde{f}(y). \tag{4.16}$$

Proof. The double limit follows from Theorem 4.2.4 and Lemma 4.4.3. If $-1 \in \mathcal{D}(C_0)$, then $\widetilde{\theta}(0)$ exists and is finite, hence Equation 4.16 follows from Equations 4.15, 4.7 and 4.6.   □

4.4.6 Note

Note that Equation 4.16 gains plausibility by the following approximation:

$$\sum_{k \in K_n(y,\epsilon)} \mu^m[B_{\delta_n}(k)] \approx \delta_n^{my} \# K_n(y,\epsilon),$$

where $K_n(y, \epsilon)$ is given in Definition 2.5.4. Then,

$$\lim_{\epsilon \to 0} \lim_{n \to \infty} \frac{\log \nu_m^{(n)}\{B_{\delta_n} : -\epsilon < Y_n - y \le \epsilon\}}{-\log \delta_n}$$

$$= \lim_{\epsilon \to 0} \lim_{n \to \infty} \frac{1}{-\log \delta_n} \log \left(\frac{\sum_{k \in K_n(y,\epsilon)} \mu^m[B_{\delta_n}(k)]}{\sum_{k \in K_n} \mu^m[B_{\delta_n}(k)]} \right)$$

$$\approx \lim_{\epsilon \to 0} \lim_{n \to \infty} \frac{1}{-\log \delta_n} \log \left(\delta_n^{my} \# K_n(y, \epsilon) \right) + \tilde{\theta}(m)$$

$$= -my + \tilde{f}(y) + \tilde{\theta}(m).$$

$\square$

4.4.7 Corollary

Given that $C_0(q)$ satisfies the Extended Hypotheses 4.2.2 and $-m \in \mathrm{int}\,\mathcal{D}(C_0)$, then $y_m = C'_m(0) = C'_0(-m)$ is the unique value of y at which the infimum is attained in $\tilde{\theta}(m) = \inf_y \{my - \tilde{f}(y)\}$. Further, y_m is a continuous function of m for all m where $-m \in \mathrm{int}\,\mathcal{D}(C_0)$.

Proof. Continuity of y_m follows from the hypothesis that $C_0(q)$ is differentiable on $\mathrm{int}\,\mathcal{D}(C_0)$. From Corollary 4.4.4,

$$\inf_y I_m(y) = I_m(y_m) = 0,$$

where y_m is unique. From Equation 4.16,

$$\inf_y I_m(y) = \inf_y \left\{ -\tilde{\theta}(m) + my - \tilde{f}(y) \right\} = 0,$$

hence the result. $\square$

4.4.8 Corollary

Given that $C_0(q)$, or equivalently $\tilde{\theta}(0) - \tilde{\theta}(-q)$, satisfies the Extended Hypotheses 4.2.2 and $-1 \in \mathcal{D}(C_0)$, then the measure μ is a multifractal measure satisfying the lattice formalism in a weak sense (Definition 2.5.4). Further, the value of y where $\tilde{\theta}(q) = \inf_y \{qy - \tilde{f}(y)\}$ occurs at $y_q = \tilde{\theta}'(-q)$ for $-q \in \mathrm{int}\,\mathcal{D}(C_0)$.

Proof. The first statement follows from Corollary 4.3.5. Solution of the infimum follows from Equation 4.6 and Corollary 4.4.7. $\square$

4.4.9 Example

Consider the same situation as in Example 4.3.3, but where the sampling measure is $\nu_m^{(n)}$. In this case there is a series of limit points y_m, where

$$y_m = C'_m(0) = \frac{-1}{\left(\sum_{i \in \Omega_0} p_i^q\right)} \sum_{\omega \in \Omega_0} p_\omega^q \log_b p_\omega,$$

the same as y_q derived in Chapter 3, where Lagrange multipliers were used.   □

The results so far are somewhat analogous to a weak law of large numbers (see Example 4.3.3). We have shown that under certain conditions, given by Extended Hypotheses 4.2.2, $\widetilde{\theta}(q)$ and $\widetilde{f}(y)$ are related via Legendre transforms. That is, global averaging is related to the local behaviour of the measure μ. The multifractal spectrum $\widetilde{f}(y)$ is a dimension in the sense that it is a powerlaw exponent of the probability distribution of the local behaviour of the measure μ. However, it is not necessarily a geometrical dimension (e.g., Hausdorff, box, packing) of some predefined set. The interpretation of $\widetilde{f}(y)$ as a Hausdorff dimension depends on the extension of $\nu_m^{(n)}$ to $\mathcal{B}(\mathcal{X})$. In this sense, results relating to Hausdorff dimensions are more analogous to a strong law of large numbers. We develop these ideas in the next section.

4.5 Hausdorff Dimensions

We have denoted $B_{\delta_n}(k)$ as the kth box in the lattice covering. The measure $\nu_n^{(m)}$ was defined on $\mathcal{F}_n$, the sub-σ-field generated by boxes at the nth stage. Now consider a point $x \in \mathcal{X}$. We will be interested in the box that contains this point. It is contained by the box with index $k(x)$, i.e., the box $B_{\delta_n}(k(x))$. To avoid notation becoming too clumsy, we will simply denote this as $B_{\delta_n}(x)$, if in the given context, it cannot be confused with $B_{\delta_n}(k)$. Further, let $\nu_m^{(\infty)}$ be the inductive limit measure of $\nu_m^{(n)}$ as $n \to \infty$.

4.5.1 Definition

For $y > 0$, define the sets $\widetilde{F}(y)$ and $\widetilde{F}_m(y)$ as

$$\widetilde{F}(y) = \left\{x \in \mathcal{X} : \lim_{n \to \infty} Y_n(x) = y\right\} = \left\{x \in \mathcal{X} : \lim_{n \to \infty} \frac{\log \mu[B_{\delta_n}(x)]}{\log \delta_n} = y\right\},$$

(4.17)

and

$$\widetilde{F}_m(y) = \left\{x \in \mathcal{X} : \lim_{n \to \infty} \frac{\log \nu_m^{(\infty)}[B_{\delta_n}(x)]}{\log \delta_n} = y\right\}.$$

□

In this section we evaluate the Hausdorff dimension of $\widetilde{F}(y)$. This involves a couple of steps. The argument is based on the behaviour of the sampling measure $\nu_m^{(n)}$ defined in §4.4. However, here we require the extension of that measure to $\mathcal{B}(\mathcal{X})$. As a consequence, $Y_n \to y_m = C'_0(-m)$ $\nu_m^{(\infty)}$-a.s., and hence we have something more akin to a strong law of large numbers. This is useful, because it can then be shown that the sets $\widetilde{F}(y_m)$ and $\widetilde{F}_m(my_m - \widetilde{\theta}(m))$ are the same, apart from a set of measure zero. Therefore, we can study the local behaviour of $\nu_m^{(\infty)}$. We then apply a theorem due to Billingsley to determine the Hausdorff dimension of $\widetilde{F}_m(my_m - \widetilde{\theta}(m))$. This is then related back to the set of interest, $\widetilde{F}(y)$.

4.5.2 Extension Theorem

Assume that $\nu_m^{(n)}$ is consistent, i.e.,

$$\nu_m^{(n+1)}(E) = \nu_m^{(n)}(E) \qquad \forall E \in \mathcal{F}_n \text{ and } n = 1, 2, \cdots,$$

and

$$\nu_m^{(\infty)}(E) = \nu_m^{(n)}(E) \qquad \forall E \in \mathcal{F}_n \text{ and } n = 1, 2, \cdots,$$

where $\nu_m^{(\infty)}$ is the inductive limit measure of $\nu_m^{(n)}$ as $n \to \infty$. Then $\nu_m^{(\infty)}$ is a unique extension of $\nu_m^{(n)}$ to $\mathcal{B}([0, 1]^d)$.

Proof. If $E \in \mathcal{F}_n \Rightarrow E \in \mathcal{F}_{n+1}$. The result follows from the Kolmogorov extension theorem (Rényi, 1970, page 286; Breiman, 1968, page 24). $\qquad \square$

4.5.3 Lemma

Assume that $\nu_m^{(n)}$ is consistent as in Theorem 4.5.2. Then $Y_n \to y_m$ $\nu_m^{(\infty)}$-a.s., or equivalently $\nu_m^{(\infty)}\big[\widetilde{F}(y_m)\big] = 1$.

Proof. It follows from Theorem 4.5.2 that $\nu_m^{(\infty)}$ is a unique extension of $\nu_m^{(n)}$ to $\mathcal{B}(\mathcal{X})$. Hence, from Theorem B.3.15, $Y_n \to y_m$ $\nu_m^{(\infty)}$-a.s. $\qquad \square$

4.5.4 Lemma

Assume that $\nu_m^{(n)}$ is consistent as in Theorem 4.5.2. Then

$$\widetilde{F}(y_m) \subseteq \widetilde{F}_m\left(my_m - \widetilde{\theta}(m)\right).$$

Further, that part where there is no intersection has measure zero, i.e.,

$$\nu_m^{(\infty)}\left[\widetilde{F}_m(my_m - \widetilde{\theta}(m)) \setminus \widetilde{F}(y_m)\right] = 0.$$

Proof. For all $x \in \widetilde{F}(y_m)$

$$\lim_{n \to \infty} \frac{\log \nu_m^{(\infty)}[B_{\delta_n}(x)]}{\log \delta_n} = \lim_{n \to \infty} \frac{\log \nu_m^{(n)}[B_{\delta_n}(x)]}{\log \delta_n}$$

$$= \lim_{n \to \infty} \left(\frac{\log \mu^m[B_{\delta_n}(x)]}{\log \delta_n} - \frac{\sum_k \log \mu^m[B_{\delta_n}(k)]}{\log \delta_n}\right)$$

$$= my_m - \widetilde{\theta}(m).$$

Thus $\widetilde{F}(y_m) \subseteq \widetilde{F}_m(my_m - \widetilde{\theta}(m))$. The second statement follows from Lemma 4.5.3. $\qquad\qquad\square$

We need a result due to Billingsley (1965). We initially give a generalisation of the Hausdorff dimension, then Billingsley's theorem.

4.5.5 Generalised Hausdorff Dimension Definition (Cutler, 1986)

Suppose that $(\mathcal{X}, \mathcal{B}(\mathcal{X}), \nu)$ is a probability space where $\mathcal{X} \subseteq \mathbb{R}^d$ and $\mathcal{B}(\mathcal{X})$ contains the Borel sets of $\mathcal{X}$. Then define the *Generalised s-Dimensional Hausdorff Measure* of $F \subseteq \mathcal{X}$ as

$$\mathcal{G}^s(F) = \lim_{\delta \to 0} \inf_{\{U_i\}} \left\{ \sum_{i=1}^{\infty} \nu(U_i)^s : \nu(U_i) \leq \delta \text{ and } F \subseteq \bigcup_{i=1}^{\infty} U_i \right\}.$$

The *Generalised Hausdorff Dimension* with respect to the measure ν, $\dim_\nu(F)$, is

$$\dim_\nu(F) = \inf\{s : \mathcal{G}^s(F) = 0\} = \sup\{s : \mathcal{G}^s(F) = \infty\}.$$

$\qquad\qquad\square$

In what follows, we use lattice coverings by b-adic cubes. A lattice covering of d-dimensional b-adic cubes at the nth level has boundaries determined by the b-adic rational expansions of length n, but where each boundary is open on the right.

4.5.6 Billingsley's Theorem (Billingsley, 1965; Cutler, 1986, Page 1477)

If λ and ν are non-atomic probability measures on $\mathcal{B}([0, 1]^d)$ and

$$G(y) \subseteq \left\{ x : \lim_{n \to \infty} \frac{\log \nu[B_n(x)]}{\log \lambda[B_n(x)]} = y \right\},$$

where $B_n(x)$ is the d-dimensional b-adic cube of volume b^{-nd} containing x, then

$$\dim_\lambda G(y) = y \dim_\nu G(y).$$

$\qquad\qquad\square$

4.5.7 Corollary (Billingsley, 1965; Cutler, 1986, Page 1477)

Assume that λ and ν are non-atomic probability measures on $\mathcal{B}([0, 1]^d)$, $G(y)$ is as in Theorem 4.5.6, and $\nu[G(y)] > 0$. Then $\dim_\nu G(y) = 1$. Further, let λ denote the Lebesgue measure. Then

$$\dim_H G(y) = d \dim_\lambda G(y) = dy.$$

$\qquad\qquad\square$

4.5.8 Lemma

Assume that $\nu_m^{(n)}$ is consistent as in Theorem 4.5.2. Let $\nu_m^{(\infty)}$, the inductive limit of $\nu_m^{(n)}$, be a non-atomic probability measure on $\mathcal{B}([0,1]^d)$. Then,

$$\dim_H \widetilde{F}_m \left(m y_m - \widetilde{\theta}(m) \right) = m y_m - \widetilde{\theta}(m).$$

Proof. In Corollary 4.5.7, let $\nu = \nu_m^{(\infty)}$. From Lemma 4.5.4,

$$\nu_m^{(\infty)} \left[\widetilde{F}_m (m y_m - \widetilde{\theta}(m)) \right] = 1.$$

The result follows by noting that $G(y/d) = \widetilde{F}_m(y)$. $\qquad\qquad\square$

4.5.9 Theorem

Let $\mathcal{X} = [0,1]^d$ and $B_{\delta_n}(k)$ form a system of b-adic cubes of width $\delta_n = b^{-n}$, where b is an integer ≥ 2. Given $C_0(q)$ satisfies the Extended Hypotheses 4.2.2, $-m \in \mathrm{int}\mathcal{D}(C_0)$ and $-1 \in \mathcal{D}(C_0)$, and $\nu_m^{(n)}$ is consistent as in Theorem 4.5.2, then

$$\dim_H \widetilde{F}(y_m) = \widetilde{f}(y_m),$$

where $y_m = C_m'(0) = C_0'(-m)$.

Proof. From Lemmas 4.5.4 and 4.5.8, $\dim_H \widetilde{F}(y_m) = m y_m - \widetilde{\theta}(m)$. From Corollary 4.4.4, $I_m(y_m) = 0$, and hence the result follows from Equation 4.16. $\qquad\square$

4.5.10 Corollary

Let $\mathcal{X} = [0,1]^d$ and $B_{\delta_n}(k)$ form a system of b-adic cubes of width $\delta_n = b^{-n}$, where b is an integer ≥ 2. Given that $C_0(q)$ satisfies the Extended Hypotheses 4.2.2, $-1 \in \mathcal{D}(C_0)$, and for all m such that $-m \in \mathcal{D}(C_0)$, $\nu_m^{(n)}$ is consistent as in Theorem 4.5.2, then

$$\dim_H \widetilde{F}(y) = \widetilde{f}(y) \qquad \text{for } y \in (y_{\min}, y_{\max}),$$

where $y_{\min} = \lim_{m \to q_{\max}} y_m$, $y_{\max} = \lim_{m \to q_{\min}} y_m$, $(q_{\min}, q_{\max}) = \mathcal{D}(C_0)$ and $y_m = C_0'(-m)$.

Proof. From Theorem 4.5.9, $\dim_H \widetilde{F}(y_m) = \widetilde{f}(y_m)$. The result follows because $y_m = C_0'(-m)$ is a continuous function of m taking all values on $(y_{\min}, y_{\max})$. $\qquad\square$

4.5.11 Example - Multinomial Measures

The multinomial measures of Chapter 3 fit into this framework, with $\mathcal{X} = [0,1]$ and $\delta_n = b^{-n}$ where b is a positive integer ≥ 2. In this chapter the boxes were denoted by $B_{\delta_n}(k)$ where k is some arbitrary index. In this example it is easier to denote the boxes by $J_n(\omega)$ where $\omega \in \Omega^n$, a space of sequences of length n (see Chapter 3). However, the lattice system $J_n(\omega)$, $\omega \in \Omega^n$, refers to the same lattice system as the $B_{\delta_n}(k)$'s, except that $J_n(\omega)$ is a closed interval, and $B_{\delta_n}(k)$ is half open. By specifying $B_{\delta_n}(k)$ to be half open ensures that each $x \in \mathcal{X}$ belongs to a unique box, and hence $\widetilde{F}(y)$ in Equation 4.17 is well defined. In the case of the multinomial measures, the b-adic expansion of a number x allocates it to a unique box. In the case of the multinomial measures, the boundaries of the intervals have zero measure, and hence the functions $\widetilde{\theta}(q)$ and $\widetilde{f}(y)$ will be the same under either covering regime.

In the current example, $\mathcal{F}_n$ could also be defined as the sub-σ-field generated by the non-overlapping intervals $J_n(\omega)$, $\omega \in \Omega^n$. Thus if $A \in \mathcal{F}_n$, then $A \cap J_n(\omega) = \emptyset$ or $A \cap J_n(\omega) = J_n(\omega)$. The measure μ is defined on $\mathcal{F}_n$ by Equation 3.1. By construction, it is clear that it has an extension to $\mathcal{B}([0,1])$.

Now consider the family of sampling measures. Recall that $\Omega_0 = \{\omega \in \Omega : p_\omega > 0\}$ and hence $\Omega_0^n = \{\omega \in \Omega^n : \mu[J_n(\omega)] > 0\}$. Let $A \in \mathcal{F}_n$. Then $A \in \mathcal{F}_{n+1}$, hence

$$\nu_m^{(n+1)}(A) = \frac{\sum_{\omega \in \Omega_0^{n+1}} M_m[A \cap J_{n+1}(\omega)]}{(\sum_{\omega \in \Omega_0} p_\omega^m)^{n+1}}$$

$$= \frac{\sum_{\omega \in \Omega_0^n} \sum_{\sigma \in \Omega_0} M_m[A \cap J_{n+1}(\omega \bowtie \sigma)]}{(\sum_{\omega \in \Omega_0} p_\omega^m)^{n+1}}$$

where $\omega \bowtie \sigma$ is to be interpreted as the concatenation of the digit σ to the end of the n length sequence ω and the set function M_m is defined by Equation 4.12. Recall that $A \in \mathcal{F}_n$, hence for a given $\omega \in \Omega_0^n$, if $A \cap J_{n+1}(\omega \bowtie \sigma) \neq \emptyset$ for a given $\sigma \in \Omega_0$, then $A \cap J_{n+1}(\omega \bowtie \sigma) \neq \emptyset$ for all $\sigma \in \Omega_0$. Therefore,

$$\nu_m^{(n+1)}(A) = \frac{\sum_{\omega \in \Omega_0^n} M_m[A \cap J_n(\omega)] \sum_{\sigma \in \Omega_0} p_\sigma^m}{(\sum_{\omega \in \Omega_0} p_\omega^m)^{n+1}}$$

$$= \nu_m^{(n)}(A).$$

Hence $\nu_m^{(n)}$ is consistent. By Theorem 4.5.2, $\nu_m^{(n)}$ has an extension to $\mathcal{B}([0,1])$ denoted by $\nu_m^{(\infty)}$. From Equation 3.8,

$$\widetilde{\theta}(q) = -\log_b \sum_{\omega \in \Omega_0} p_\omega^q,$$

and from Equation 4.6, $C_0(q) = \widetilde{\theta}(0) - \widetilde{\theta}(-q)$. Hence

$$I_0(y) = \sup_q \{qy - C_0(q)\} = \widetilde{D}_0 - \inf_q \left\{ qy - \widetilde{\theta}(q) \right\}.$$

Given continuity of $I_0(y)$, then $I_0(y) = \widetilde{D}_0 - \widetilde{f}(y)$, and further $\dim_H \widetilde{F}(y) = \widetilde{f}(y)$. The Legendre transform relationships in Equations 4.9 and 4.10 are shown graphically in Figures 3.2, 3.3 and 3.4. $\qquad \square$

CHAPTER 5

Point Centred Multifractals

5.1 Introduction

In this chapter we follow a parallel development to that in Chapter 4, though use a point centred construction instead. That is, we are interested in the relationship between the functions $f(y)$ and $\theta(q)$ defined by Equations 2.9 and 2.5, respectively, and when $f(y)$ can be interpreted as a Hausdorff dimension.

In Chapter 4, the behaviour of the measure was analysed by covering the set $\mathcal{X} \subseteq \mathbb{R}^d$ by d-dimensional boxes of width δ_n, where $\delta_n \to 0$ as $n \to \infty$. A sub-σ-field, $\mathcal{F}_n$, was then generated by the lattice boxes at the nth stage. Relationships were then derived by defining a family of sampling measures on $\mathcal{F}_n$. In this chapter, a family of sampling measures are also defined, but all members of the family are on the same measure space, i.e., $(\mathcal{X}, \mathcal{B}(\mathcal{X}))$.

In §5.2, the large deviation formalism is defined for a general sampling measure. In §5.3, a family of sampling measures is defined, and in §5.4 conditions are given under which $f(y)$ can be interpreted as a Hausdorff dimension. These three sections closely follow the line of argument in Chapter 4. Less explanation is given in this chapter, as it is assumed that the reader has already studied the material in Chapter 4.

In §5.5, relationships between the lattice and point centred constructions are investigated. Of particular interest are the questions: when are the Rényi dimensions and multifractal spectrums the same in both constructions. Another difference between the two constructions is the nature and existence of the limits in δ, and the width of the covering lattices or spheres. These will also be discussed in §5.5.

5.2 Large Deviation Formalism

We base our construction on the probability space $(\mathcal{X}, \mathcal{B}(\mathcal{X}), \nu^{(\delta)})$ where $\mathcal{B}(\mathcal{X})$ are the Borel sets of $\mathcal{X} \subseteq \mathbb{R}^d$, and $\nu^{(\delta)}$ is some general *sampling* measure, which is a function of δ. Note the difference with the construction based on lattices in §4.2, where we had a sequence of probability spaces $(\mathcal{X}, \mathcal{F}_n, \nu^{(n)})$. Here we have the one measure space $(\mathcal{X}, \mathcal{B}(\mathcal{X}))$, but with the probability measure $\nu^{(\delta)}$ changing with δ.

Define the measurable functions $\{U_\delta\}_{\delta>0}$ where

$$U_\delta : (\mathcal{X}, \mathcal{B}(\mathcal{X})) \to (\mathbb{R}, \mathcal{B}(\mathbb{R})),$$

as

$$U_\delta(x) = \begin{cases} -\log \mu[S_\delta(x)] & \text{if } \mu[S_\delta(x)] > 0 \\ 0 & \text{if } \mu[S_\delta(x)] = 0, \end{cases}$$

where $S_\delta(x)$ is a d-dimensional closed sphere of radius δ centred at x. Also define the functions $\{Y_\delta\}_{\delta>0}$ as

$$Y_\delta(x) = \frac{U_\delta(x)}{-\log \delta}.$$

We are interested in the behaviour of Y_δ on the probability space $(\mathcal{X}, \mathcal{B}(\mathcal{X}), \nu^{(\delta)})$ as $\delta \to 0$.

Let $\mathcal{X}_\delta = \{x \in \mathcal{X} : \mu[S_\delta(x)] > 0\}$. $\mathcal{X}_\delta$ is sometimes referred to as the δ-parallel body of the support of μ or a sausage set. It follows from Falconer (1990, Proposition 3.2) that

$$\widetilde{D}_0 = \dim_B \operatorname{supp}(\mu) = d - \lim_{\delta \to 0} \frac{\log \operatorname{Vol}^d(\mathcal{X}_\delta)}{\log \delta}$$

if the limit exists, and $\operatorname{Vol}^d(\mathcal{X}_\delta)$ is the d-dimensional volume of $\mathcal{X}_\delta$.

Note that all references to $C(q)$, $I(y)$, $C_m(q)$ and $I_m(y)$ are to those defined within this chapter, and do not refer to those in Chapter 4.

5.2.1 Rescaled Cumulant Generating Function

Using Equation B.5, it follows that the *Rescaled Cumulant Generating Function* is $C(q)$, $q \in \mathbb{R}$, if the limit ($\delta \to 0$) exists, where

$$C(q) = \lim_{\delta \to 0} \frac{\log \mathrm{E}\left[e^{qU_\delta(x)}\right]}{-\log \delta},$$

and the expectation is taken with respect to the sampling measure $\nu^{(\delta)}$. Therefore,

$$C(q) = \lim_{\delta \to 0} \frac{\log \int_{\mathcal{X}_\delta} \mu^{-q}[S_\delta(x)]\nu^{(\delta)}(dx)}{-\log \delta}. \tag{5.1}$$

Note that we allow $+\infty$ as a limit value. Further, let

$$I(y) = \sup_{q \in \mathbb{R}}\{qy - C(q)\}, \quad y > 0. \tag{5.2}$$

$I(y)$ may also take the value $+\infty$, and is referred to as the Legendre transform of $C(q)$. $\square$

5.2.2 Extended Hypotheses

1. $C(q)$ satisfies Hypotheses B.3.8.
2. $C(q)$ is differentiable on the interior of the $\mathcal{D}(C) = \{q \in \mathbb{R} : C(q) < \infty\}$, and $C(q)$ is steep (Definition B.2.1). $\square$

5.2.3 Theorem - Exponential Convergence

If $C(q)$ satisfies the Extended Hypotheses 5.2.2, then $I(y)$ has a unique minimum at $y_0 = C'(0)$ such that $I(y_0) = 0$. Further, $Y_\delta \xrightarrow{\text{exp}} y_0$ with respect to $\nu^{(\delta)}$ (see Definition B.3.13), and

$$C(q) = \sup_y \{qy - I(y)\}. \tag{5.3}$$

Proof. Given the Extended Hypotheses 5.2.2, the argument follows in exactly the same way as for Theorem 4.2.3.   □

5.2.4 Theorem - Large Deviation Bounds

If $C(q)$ satisfies the Extended Hypotheses 5.2.2, then

$$\lim_{\epsilon \to 0} \lim_{\delta \to 0} \frac{\log \nu^{(\delta)}\{x : -\epsilon < Y_\delta(x) - y \le \epsilon\}}{-\log \delta} = -I(y) \quad \text{for } y \in \text{int}\mathcal{D}(I).$$
$$\tag{5.4}$$

Proof. Given the Extended Hypotheses 5.2.2, the argument follows in exactly the same way as in Theorem 4.2.4.   □

5.3 A Family of Sampling Measures

A family of sampling measures can be defined in a similar manner as in §4.4 so that different limits are realised depending on the sampling measure.

5.3.1 Definition

Define a family of sampling measures $\nu_m^{(\delta)}$ as

$$\nu_m^{(\delta)}(E) = \frac{\int_{E \cap \mathcal{X}_\delta} \mu^{m-1}[S_\delta(x)]\mu(dx)}{\int_{\mathcal{X}_\delta} \mu^{m-1}[S_\delta(x)]\mu(dx)}, \tag{5.5}$$

where $m \in \mathbb{R}$, δ is fixed, and $E \in \mathcal{B}(\mathcal{X})$.   □

In this case, $C(q)$ and $I(y)$ will be denoted as $C_m(q)$ and $I_m(y)$, respectively. Note that unlike the family in §4.4 which is defined on a sub-σ-field, here $\nu_m^{(\delta)}$ is defined on $\mathcal{B}(\mathcal{X})$.

5.3.2 Theorem - Rescaled Cumulant Generating Function

If $\theta(0)$ exists, or equivalently $-1 \in \mathcal{D}(C_0)$, then

$$C_0(q) = \theta(0) - \theta(-q),$$

where $\theta(q)$ is given in Definition 2.4.1. Further, if $-m \in \mathcal{D}(C_0)$, then

$$C_m(q) = C_0(q - m) - C_0(-m), \tag{5.6}$$

and

$$I_m(y) = C_0(-m) + my + I_0(y). \qquad (5.7)$$

Proof. Inserting $\nu_m^{(\delta)}$ into Equation 5.1 and letting $m = 0$ and $q = -1$, one gets $C_0(-1) = \theta(0)$, i.e., $\theta(0)$ exists iff $-1 \in \mathcal{D}(C_0)$. Thus, $C_0(q) = \theta(0) - \theta(-q)$ if $-1 \in \mathcal{D}(C_0)$. Similarly, inserting $\nu_m^{(\delta)}$ into Equation 5.1 and rearranging gives Equation 5.6, given that $-m \in \mathcal{D}(C_0)$. Equation 5.7 follows in the same manner as in Theorem 4.4.2. □

5.3.3 Lemma

Given that $C_0(q)$ satisfies the Extended Hypotheses 5.2.2 and $-m \in \mathrm{int}\mathcal{D}(C_0)$, then $C_m(q)$ satisfies the Extended Hypotheses 5.2.2. Further, $I_m(y)$ is a continuous function on $\mathcal{D}(I_0)$.

Proof. As for Lemma 4.4.3. □

5.3.4 Corollary - Exponential Convergence

If $C_0(q)$ satisfies the Extended Hypotheses 5.2.2 and $-m \in \mathrm{int}\mathcal{D}(C_0)$, then $Y_\delta \xrightarrow{\exp} y_m$ with respect to $\nu_m^{(\delta)}$, where $y_m = C'_m(0) = C'_0(-m)$. Further, $I_m(y_m) = 0$ is the unique minimum of $I_m(y)$.

Proof. Follows directly from Theorem 5.2.3 and Lemma 5.3.3. □

5.3.5 Note

If $\theta(0)$ exists (Equation 2.5), then it follows from Equation 2.9 that

$$f(y) = -\theta(0) + \lim_{\epsilon \to 0} \lim_{\delta \to 0} \frac{\log \nu_0^{(\delta)}\{x \in \mathcal{X} : -\epsilon < Y_\delta(x) - y \le \epsilon\}}{-\log \delta}, \qquad (5.8)$$

allowing for $f(y) = -\infty$ when $\nu_0^{(\delta)}\{x \in \mathcal{X} : -\epsilon < Y_\delta(x) - y \le \epsilon\} = 0$. □

5.3.6 Corollary - Large Deviation Bounds

Given that $C_0(q)$ satisfies the Extended Hypotheses 5.2.2, then for all m such that $-m \in \mathrm{int}\mathcal{D}(C_0)$, and for all $y \in \mathrm{int}\mathcal{D}(I_0)$

$$\lim_{\epsilon \to 0} \lim_{\delta \to 0} \frac{\log \nu_m^{(\delta)}\{x \in \mathcal{X} : -\epsilon < Y_\delta(x) - y \le \epsilon\}}{-\log \delta} = -I_m(y).$$

Further, if $-1 \in \mathcal{D}(C_0)$, then

$$I_m(y) = -\theta(m) + my - f(y), \qquad (5.9)$$

where $\theta(m)$ and $f(y)$ are given by Equations 2.5 and 2.9, respectively.

Proof. The double limit follows from Theorem 5.2.4. Given $-1 \in \mathcal{D}(C_0)$, then $\theta(0)$ exists and is finite. Together with Equation 5.8, it then follows that $I_0(y) = -f(y) - \theta(0)$. Substituting into Equation 5.7 gives Equation 5.9. $\qquad\square$

5.3.7 Corollary

Given that $C_0(q)$ satisfies the Extended Hypotheses 5.2.2 and $-m \in \operatorname{int}\mathcal{D}(C_0)$, then $y_m = C_m'(0) = C_0'(-m)$ is the unique value of y at which the infimum is attained in $\theta(m) = \inf_y \{my - f(y)\}$. Further, y_m is a continuous function of m.

Proof. In the same way as for Corollary 4.4.7. $\qquad\square$

5.3.8 Corollary

Given that $C_0(q)$ satisfies the Extended Hypotheses 5.2.2 and $-1 \in \mathcal{D}(C_0)$, then the measure μ is a multifractal measure satisfying the point centred formalism in a weak sense (Definition 2.5.5). Further, the value of y where $\theta(q) = \inf_y \{qy - f(y)\}$ occurs at $y = \theta'(-q)$ for $-q \in \operatorname{int}\mathcal{D}(C_0)$.

Proof. The argument follows in the same manner as for Corollary 4.4.8. $\qquad\square$

5.4 Hausdorff Dimensions

In this section we follow a similar line of argument as was done in §4.5, where conditions were determined under which $\widetilde{f}(y)$ could be interpreted as the Hausdorff dimension of $\widetilde{F}(y)$. In §4.5, we appealed to a result by Billingsley (1965). The analogue in the present context is a result due to Young (1982).

5.4.1 Definition

For $y > 0$, define the sets $F(y)$ and $F_m(y)$ as

$$F(y) = \left\{ x \in \mathcal{X} : \lim_{\delta \to 0} Y_\delta(x) = y \right\} = \left\{ x \in \mathcal{X} : \lim_{\delta \to 0} \frac{\log \mu[S_\delta(x)]}{\log \delta} = y \right\}$$
(5.10)

and

$$F_m(y) = \left\{ x \in \mathcal{X} : \lim_{\delta \to 0} \frac{\log \nu_m^{(0^+)}[S_\delta(x)]}{\log \delta} = y \right\},$$

respectively, where $\nu_m^{(0^+)}$ is the inductive limit measure of $\nu_m^{(\delta)}$ as $\delta \to 0$. $\qquad\square$

For the argument to work, we are interested in the behaviour of Y_δ on the probability space $(\mathcal{X}, \mathcal{B}(\mathcal{X}), \nu_m^{(0^+)})$. Under satisfactory conditions, it will be shown that $F(y_m) \subseteq F_m(my_m - \theta(m))$. To do this, a few regularity conditions must be satisfied.

5.4.2 Lemma

Assume that $C_0(q)$ satisfies the Extended Hypotheses 5.2.2 and $-1 \in \mathrm{int}\mathcal{D}(C_0)$. Further, for all m such that $-m \in \mathrm{int}\mathcal{D}(C_0)$, we assume that $\nu_m^{(0^+)}[F(y_m)] = 1$ and that the following two relationships hold for all $x \in F(y_m)$:

$$\lim_{\delta \to 0} \frac{\log \int_{S_\delta(x)} \mu^{m-1}[S_\delta(z)]\mu(dz)}{\log \delta} = \lim_{\delta \to 0} \frac{\log \mu^m[S_\delta(x)]}{\log \delta} \qquad (5.11)$$

and

$$\lim_{\delta \to 0} \frac{\log \nu_m^{(0^+)}[S_\delta(x)]}{\log \delta} = \lim_{\delta \to 0} \frac{\log \nu_m^{(\delta)}[S_\delta(x)]}{\log \delta}. \qquad (5.12)$$

Then $F(y_m) \subseteq F_m(my_m - \theta(m))$ where $y_m = C'_m(0) = C'_0(-m)$. Further, that part where there is no intersection has measure zero, i.e.,

$$\nu_m^{(0^+)}\big[F_m(my_m - \theta(m)) \setminus F(y_m)\big] = 0.$$

Proof. Since $-1 \in \mathcal{D}(C_0)$, then $\theta(0)$ exists. Since also $-m \in \mathrm{int}\mathcal{D}(C_0)$, then both $\theta(m)$ and y_m exist. Now consider those $x \in F(y_m)$, then

$$\begin{aligned}
\lim_{\delta \to 0} \frac{\log \nu_m^{(0^+)}[S_\delta(x)]}{\log \delta} &= \lim_{\delta \to 0} \frac{\log \int_{S_\delta(x)} \mu^{m-1}[S_\delta(x)]\mu(dz)}{\log \delta} - \theta(m) \\
&= my_m - \theta(m).
\end{aligned}$$

Hence $F(y_m) \subseteq F_m(my_m - \theta(m))$. Since $\nu_m^{(0^+)}[F(y_m)] = 1$, the last statement follows.   □

Note the difference between the above result and Lemma 4.5.4 in the lattice based constructions. In the lattice case, given consistency and the Extension Theorem 4.5.2, $\nu_m^{(\infty)}[F(y_m)] = 1$ follows from Theorem B.3.15. In the point centred case, Equation 5.12 is a weaker condition than the consistency conditions of the Extension Theorem 4.5.2 and hence we also require that $\nu_m^{(0^+)}[F(y_m)] = 1$. Equation 5.11 is a type of smoothness condition.

5.4.3 Theorem (Young, 1982)

Let ν be a finite non-atomic Borel measure on $\mathbb{R}^d$ and $F \subset \mathbb{R}^d$ be ν-measurable and have $\nu(F) > 0$. Suppose that for every $x \in F$,

$$\alpha_1 \leq \liminf_{\delta \to 0} \frac{\log \nu[S_\delta(x)]}{\log \delta} \leq \limsup_{\delta \to 0} \frac{\log \nu[S_\delta(x)]}{\log \delta} \leq \alpha_2.$$

Then $\alpha_1 \leq \dim_H F \leq \alpha_2$.   □

5.4.4 Corollary

Let $\nu_m^{(0^+)}$ be a finite non-atomic Borel measure and $\nu_m^{(0^+)}[F_m(my_m - \theta(m))] > 0$. Then for all m such that $-m \in \text{int}\mathcal{D}(C_0)$,

$$\dim_H F_m(my_m - \theta(m)) = my_m - \theta(m).$$

Proof. Follows from Theorem 5.4.3. $\qquad\square$

5.4.5 Theorem

Assume that the conditions of Lemma 5.4.2 are satisfied, and $\nu_m^{(0^+)}$ is a non-atomic probability measure. Then $\dim_H F(y_m) = f(y_m)$, where $y_m = C_m'(0) = C_0'(-m)$.

Proof. From Lemma 5.4.2, $F(y_m) \subseteq F_m(my_m - \theta(m))$. Also by assumption, $\nu_m^{(0^+)}[F(y_m)] = 1$, and so it follows from Corollary 5.4.4 that

$$\dim_H F(y_m) = my_m - \theta(m).$$

Since $I_m(y_m) = 0$, and inserting into Equation 5.9, one gets $f(y_m) = my_m - \theta(m)$. $\qquad\square$

5.4.6 Corollary

Given that all conditions of Lemma 5.4.2 are satisfied, and $\nu_m^{(0^+)}$ is a non-atomic probability measure, then

$$\dim_H F(y) = f(y) \quad \text{for } y \in (y_{\min}, y_{\max}),$$

where $y_{\min} = \lim_{m \to q_{\max}} y_m$, $y_{\max} = \lim_{m \to q_{\min}} y_m$, $(q_{\min}, q_{\max}) = \mathcal{D}(C_0)$ and $y_m = C_0'(-m)$.

Proof. The result follows from Theorem 5.4.5 because $y_m = C_0'(-m)$ is a continuous function of m taking values on $(y_{\min}, y_{\max})$. $\qquad\square$

5.5 Relationships Between Lattice and Point Centred Constructions

Using the family of sampling measures for both lattices and point centred constructions, identical relationships between entropy functions and rescaled generating functions have been derived. Results that ensure that μ is a multifractal in a weak sense (i.e., Legendre transform relationships hold) follow directly from the large deviation results of Ellis (1984, 1985), and are analogous, in some sense, to a weak law of large numbers. In both the lattice and point centred cases, the functions $\widetilde{f}(y)$ and $f(y)$ can be interpreted as a type of dimension, in that it describes the powerlaw rate of convergence of the probability measures $\nu_m^{(n)}$ and $\nu_m^{(\delta)}$, respectively. However, the stronger results involving Hausdorff dimensions require almost sure convergence of Y_n and Y_δ and, therefore, are more analogous to a strong law of large numbers.

In this section, there are four questions we want to address:

1. Under what conditions are the partitioning sets of local behaviour the same under both constructions; i.e., when does $F(y) = \widetilde{F}(y)$ where $F(y)$ and $\widetilde{F}(y)$ are given by Equations 5.10 and 4.17, respectively?

2. Under what conditions are the Rényi dimensions the same under both constructions; i.e., when does $D_q = \widetilde{D}_q$ or $\theta(q) = \widetilde{\theta}(q)$, where $\theta(q)$ and $\widetilde{\theta}(q)$ are given by Equations 2.5 and 2.2, respectively?

3. The difference between the nature of the limits in δ between the lattice and point centred constructions, i.e., a sequence $\delta_n \to 0$ compared to $\delta \to 0$.

4. Under what conditions are the multifractal spectrums the same under both constructions; i.e., when does $f(y) = \widetilde{f}(y)$, where $f(y)$ and $\widetilde{f}(y)$ are given by Equations 2.9 and 2.7, respectively?

None of the questions are answered conclusively, though examples are given where these relationships do hold. These may be suggestive of the required conditions in more general situations.

When Does $F(y) = \widetilde{F}(y)$?

The following results are special cases of those in Cawley & Mauldin (1992) applied to the multinomial measures of Example 4.5.11. They also hold in the more general setting of Moran cascades to be discussed in Chapter 6. Notation used below (i.e., Ω, Ω_0, Ω^n, Ω^∞, etc.) is the same as in Example 4.5.11. Further, if $\omega = (\omega_1, \omega_2, \cdots) \in \Omega^\infty$, then $\omega|n$ is to be interpreted as $(\omega_1, \omega_2, \cdots, \omega_n)$; and the symbol $\bowtie$ is to be interpreted as concatenate, for example $(\omega_1, \cdots, \omega_n) \bowtie t = (\omega_1, \cdots, \omega_n, t)$.

5.5.1 Theorem (Cawley & Mauldin, 1992)

Let $\omega = (\omega_1, \omega_2, \cdots) \in \Omega^\infty$,

$$x = \sum_{k=1}^{\infty} \frac{\omega_k}{b^k},$$

and $S_\delta(x)$ be a closed sphere of radius δ about x. Further, let μ be a multinomial measure of base b (see Example 4.5.11). Then,

$$\limsup_{\delta \to 0} \frac{\log \mu[S_\delta(x)]}{\log \delta} \leq \limsup_{n \to \infty} \frac{\log \mu[J_n(\omega|n)]}{\log \delta_n},$$

where $\delta_n = b^{-n}$ and

$$J_n(\omega|n) = \left[\sum_{j=1}^{n} \frac{\omega_j}{b^j}, \ \frac{1}{b^n} + \sum_{j=1}^{n} \frac{\omega_j}{b^j} \right].$$

Proof. Temporarily fix $\delta < 1/(2b)$. Now select n such that $J_n(\omega|n) \subseteq S_\delta(x)$ and $J_{n-1}(\omega|(n-1)) \not\subseteq S_\delta(x)$. Thus, $|J_n(\omega|n)| \leq 2\delta$. The least optimal covering of $J_{n-1}(\omega|(n-1))$ by $S_\delta(x)$ will be when $\omega = (\omega_1, \omega_2, \cdots, \omega_{n-1}, 0, 0, \cdots)$, i.e., the centre of the sphere, x, is on the lower bound of both $J_{n-1}(\omega|(n-1))$ and $J_n(\omega|n)$. By construction, $J_{n-1}(\omega|(n-1)) \not\subseteq S_\delta(x)$, hence $|J_{n-1}(\omega|(n-1))| \geq \delta$. Hence

$$\delta \leq |J_{n-1}(\omega|(n-1))| = b|J_n(\omega|n)| \leq 2b\delta < 1.$$

Therefore,

$$
\begin{aligned}
\frac{\log \mu[S_\delta(x)]}{\log \delta} &\leq \frac{\log \mu[J_n(\omega|n)]}{\log \delta} \\
&\leq \frac{\log \mu[J_n(\omega|n)]}{\log |J_n(\omega|n)| + \log b} \\
&= \frac{\log \mu[J_n(\omega|n)]}{\log |J_n(\omega|n)| \left(1 + \dfrac{\log b}{\log |J_n(\omega|n)|}\right)} \\
&\leq \frac{\log \mu[J_n(\omega|n)]}{\log |J_n(\omega|n)|} \left(1 - \frac{\log b}{\log |J_n(\omega|n)|}\right) \\
&\leq \frac{\log \mu[J_n(\omega|n)]}{\log |J_n(\omega|n)|} \left(1 - \frac{\log b}{\log(2\delta)}\right).
\end{aligned}
$$

Taking limits gives the result.   □

5.5.2 Theorem (Cawley & Mauldin, 1992)

Let $\omega = (\omega_1, \omega_2, \cdots) \in \Omega^\infty$,

$$x = \sum_{k=1}^{\infty} \frac{\omega_k}{b^k},$$

and $S_\delta(x)$ be a closed sphere of radius δ about x. Further, let μ be a multinomial measure of base b (see Example 4.5.11), constructed with the elementary weights $p_0, \cdots, p_{b-1}$. Assume that for each $p_i > 0$ and $p_j > 0$ such that $i < j$, there exists a k such that $i < k < j$ and $p_k = 0$. Then

$$\liminf_{\delta \to 0} \frac{\log \mu[S_\delta(x)]}{\log \delta} \geq \liminf_{n \to \infty} \frac{\log \mu[J_n(\omega|n)]}{\log \delta_n},$$

where $\delta_n = b^{-n}$.

Proof. Let $\mathcal{K} = \bigcap_{n=1}^{\infty} \bigcup_{\omega' \in \Omega_0^n} J_n(\omega')$, where $\bigcup_{\omega' \in \Omega_0^n} J_n(\omega')$ are those subintervals at the nth level with positive μ measure. Temporarily fix $\delta < 1/(2b)$. Let

$$h_\delta(x) = \max\{n : S_\delta(x) \cap \mathcal{K} \subseteq J_n(\omega|n)\},$$

where $\omega = (\omega_1, \omega_2, \cdots)$. Thus $\log \mu[S_\delta(x)] \leq \log \mu[J_{h_\delta(x)}(\omega|h_\delta(x))]$. One can select a $t \in \Omega_0$ such that $t \neq \omega_{h_\delta(x)+1}$ (i.e., not equal to the $(h_\delta(x)+1)$st digit in the sequence ω) and a point $y \in J_{h_\delta(x)+1}((\omega|h_\delta(x)) \bowtie t) \cap \mathcal{K} \cap S_\delta(x)$ such that $x \in J_{h_\delta(x)+1}(\omega|(h_\delta(x)+1)) \subseteq J_{h_\delta(x)}(\omega|h_\delta(x))$ and

$$J_{h_\delta(x)+1}(\omega|(h_\delta(x)+1)) \cap J_{h_\delta(x)+1}((\omega|h_\delta(x)) \bowtie t) = \emptyset.$$

Therefore,

$$\frac{1}{b} > 2\delta > |x - y|.$$

However, there is a gap of width $b^{-(h_\delta(x)+1)}$ between the b-adic interval containing x and that containing y. Thus, $2\delta > |x - y| > b^{-(h_\delta(x)+1)} = b^{-1}|J_{h_\delta(x)}(\omega|h_\delta(x))|$. Therefore, $|J_{h_\delta(x)}(\omega|h_\delta(x))| < 2\delta b < 1$, and so

$$
\begin{aligned}
\frac{\log \mu[S_\delta(x)]}{\log \delta} &\geq \frac{\log \mu[J_{h_\delta(x)}(\omega|h_\delta(x))]}{\log \delta} \\
&\geq \frac{\log \mu[J_{h_\delta(x)}(\omega|h_\delta(x))]}{\log |J_{h_\delta(x)}(\omega|h_\delta(x))| - \log(2b)} \\
&= \frac{\log \mu[J_{h_\delta(x)}(\omega|h_\delta(x))]}{\log |J_{h_\delta(x)}(\omega|h_\delta(x))| \left(1 - \dfrac{\log(2b)}{\log |J_{h_\delta(x)}(\omega|h_\delta(x))|}\right)} \\
&\geq \frac{\log \mu[J_{h_\delta(x)}(\omega|h_\delta(x))]}{\log |J_{h_\delta(x)}(\omega|h_\delta(x))|} \left(1 + \frac{\log(2b)}{\log |J_{h_\delta(x)}(\omega|h_\delta(x))|}\right).
\end{aligned}
$$

Taking limits gives the result.   □

5.5.3 Corollary

Let μ be a multinomial measure (Example 4.5.11) satisfying the conditions of Theorem 5.5.2, and $B_{\delta_n}(x)$ be the b-adic interval of length $\delta_n = b^{-n}$ that contains the point x. Then $x \in F(y)$ iff $x \in \widetilde{F}(y)$ where $F(y)$ and $\widetilde{F}(y)$ are defined by Equations 4.17 and 5.10, respectively. Therefore, $\dim_H F(y) = \dim_H \widetilde{F}(y) = \widetilde{f}(y)$.

Proof. $F(y) = \widetilde{F}(y)$ follows from Theorems 5.5.1 and 5.5.2. $\dim_H F(y) = \dim_H \widetilde{F}(y) = \widetilde{f}(y)$ follows from Example 4.5.11.   □

Note that the above example requires 'gaps' (see Theorem 5.5.2). Note also, in the point centred case, that the limit holds for all sequences ($\delta \to 0$), whereas in the lattice case, a particular sequence $\{\delta_n\}$ was required.

When Does $D_q = \widetilde{D}_q$?

In Example 3.5.1 it was shown that $D_2 = \widetilde{D}_2$, or equivalently $\theta(2) = \widetilde{\theta}(2)$, for the Cantor measure. The argument implemented there can also be used for $q = 3, 4, \cdots$.

5.5.4 Theorem

Let μ be a multinomial measure with gaps as in Theorem 5.5.2. Then $D_q = \tilde{D}_q$ or equivalently $\theta_q = \tilde{\theta}_q$ for $q = 2, 3, \cdots$. Further, the limit $(\delta \to 0)$ in $\theta(q)$ holds for all sequences.

Proof. The argument follows as in Example 3.5.1. Let $X_1, X_2, \cdots, X_q$ be independent random variables drawn from the distribution (μ_n) on $\mathcal{K}_n$ (see definitions in Example 3.5.1). Define

$$Y_n = \max\{|X_1 - X_q|, |X_2 - X_q|, \cdots, |X_{q-1} - X_q|\},$$

then from §2.4.5,

$$\int_{\mathcal{X}_n} \mu_n[S_\delta(x)]\mu_n(dx) = \Pr\{Y_n \leq \delta\}.$$

When $n = 0$, μ_0 is just the uniform density on the unit interval. As in Example 3.5.1, $g_0(y)$ is the density of Y_0 when $X_1, \cdots, X_q$ are sampled from the uniform distribution. The actual form of $g_0(y)$ is not required in the argument. The probability distribution after the first division is given by μ_1, that is, the first subinterval has p_0 of the probability, the second p_1, etc. However, at least every alternate subinterval has zero mass, i.e., we have gaps which are at least the width of the subintervals. Thus, if $Y_1 < b^{-1}$, then *all* q points $X_1, \cdots, X_q$ must be sampled from the same subinterval. Since each subinterval has a uniform density with total mass equal to one of $p_0, \cdots, p_{b-1}$, then

$$g_1(y) = b(p_0^q + \cdots + p_{b-1}^q)g_0(by),$$

where $0 \leq y \leq b^{-1}$. The same argument is repeated at each division and, in general,

$$g_n(y) = b(p_0^q + \cdots + p_{b-1}^q)g_{n-1}(by),$$

where $0 \leq y \leq b^{-n}$. Therefore, the probability distribution function satisfies the recurrence relation

$$G_n(y) = (p_0^q + \cdots + p_{b-1}^q)G_{n-1}(by) = (p_0^q + \cdots + p_{b-1}^q)^n G_0(b^n y),$$

when $0 \leq y \leq b^{-n}$. The argument then follows in exactly the same manner as in Example 3.5.1. □

It is interesting to note that both the proofs of Theorem 5.5.4 and Corollary 5.5.3 require the measure to be supported by a set with gaps. While this condition is sufficient, it is not clear to us that it is absolutely necessary.

Mikosch & Wang (1993, Proposition 2.1) give conditions under which $\theta(q) = \tilde{\theta}(q)$ for $q > 1$ in a more general setting. They require the measure μ to have a compact concentration set and to be continuous on that set.

Riedi (1995, Proposition 20) has shown that if μ is an arbitrary Borel measure, then for any $q > 1$, one has

$$D_q = \frac{\tilde{\theta}_R(q)}{q-1},$$

where $\tilde{\theta}_R(q)$ is defined in §2.6.3. Recall that there are two differences between $\tilde{\theta}_R(q)$ and $\tilde{\theta}(q)$: $\tilde{\theta}_R(q)$ is defined on a system of overlapping lattices (i.e., a moving 3×3 grid centred on the middle box), and the limit in δ_n is replaced by $\liminf \delta \to 0$. Moreover, Riedi (1995) shows that D_q exists iff $\tilde{\theta}_R(q)$ is grid regular. $\tilde{\theta}_R(q)$ is grid regular if it exists with $\liminf \delta \to 0$ replaced by $\lim \delta \to 0$. This does not seem unreasonable as both the point centred construction and Riedi's construction, based on overlapping boxes, have a built in smoothing operation.

Nature and Existence of Limits in δ

In both the Rényi dimensions (global average) and the multifractal spectrum (local behaviour), we considered constructions that used coverings by lattices or spheres, each of a given width which tends to zero. In the point centred case, coverings by spheres were used where the radii δ tended to zero. The limit was assumed to exist for all sequences. In the case of the lattices, such a general limiting behaviour does not usually exist and a specific sequence $\{\delta_n\}$ is required.

Reidi (1995) showed (see Example 2.6.2) that in the case of the Cantor measures, $\tilde{\theta}(q)$ does not exist for $q < 1$ if the limit in the definition of $\tilde{\theta}(q)$ is changed from having $\delta_n \to 0$ to $\delta \to 0$. However, it was shown in Theorem 5.5.4 that $\theta(q)$ does exist (i.e., with $\delta \to 0$) for the multinomial measures with gaps for $q = 2, 3, \cdots$. The proof is based on the qth interpoint difference (§2.4.5), and it is not clear to us whether these ideas can be extended to negative values of q.

When Does $f(y) = \tilde{f}(y)$?

In the situation where $\theta(q) = \tilde{\theta}(q)$, and the multifractal formalism holds in a weak sense for both lattice and point centred constructions, then $f(y) = \tilde{f}(y)$.

Note that the critical conditions for a point centred construction to hold are given by Equations 5.11 and 5.12. Equation 5.11 requires a degree of smoothness or homogeneity while Equation 5.12 requires consistent powerlaw behaviour as $\delta \to 0$. If these conditions hold for the special case of the multinomial measures as in Theorem 5.5.2, and $\theta(q)$ behaves as required, then $f(y) = \dim_H F(y)$, and hence $f(y) = \tilde{f}(y)$. Note that this is not unreasonable in this example, as these measures have gaps between the different parts, hence the integral in the definition of $f(y)$, Equation 2.9, will essentially be the number of required covering spheres (see Note 2.5.3), analogous to the number of covering boxes required in the definition of $\tilde{f}(y)$ given by Equation 2.7.

CHAPTER 6

Multiplicative Cascade Processes

6.1 Introduction

In Chapters 4 and 5, a multifractal formalism was established to describe a probability measure μ on $(\mathcal{X}, \mathcal{B}(\mathcal{X}))$, where $\mathcal{X} \subseteq \mathbb{R}^d$ and $\mathcal{B}(\mathcal{X})$ are the Borel sets of $\mathcal{X}$. Except for the multinomial measures, no further assumptions were made about the generating mechanism of μ. In this chapter, we will assume that μ is generated as the result of a cascade process.

The motivation for cascade processes comes from physics, particularly turbulence. Here energy enters the system on a very large scale, both in terms of space and the amount of energy. Then the energy is dissipated, but not in a uniform manner. Parts of the space may have eddies and quite violent behaviour, while other parts have relative calm. This space division and energy dissipation repeats down to smaller and smaller scales, until eventually, the energy is dissipated as heat. This general concept is very similar to the example of the multinomial measures in Chapter 3. However, in this chapter, there are two further generalisations that will be made. Firstly, the space may not necessarily be divided into sets of the same size at each division, the set size may even be random. Secondly, the energy dissipation or probability allocation may also be done in a random manner.

We could think of the measure space $(\mathcal{X}, \mathcal{B}(\mathcal{X}), \mu)$ as the *observation space*, where $\mathcal{X} \subseteq \mathbb{R}^d$, and μ is the so called multifractal measure. We have referred to it as a measure space because when μ is a random measure, it will not be a probability measure, though we will require that its expected value is one. In other situations, like the multinomial measures of Chapter 3, μ will be a probability measure. In this introduction, we setup a generic framework that can be used to describe both situations. Cascade processes involve defining a measure μ_n at a 'coarse' level, and then μ will be defined as the inductive limit on an infinitely 'fine' level.

The support of the measure μ is constructed in an iterative manner. At the nth stage of the process, we want a collection of non-overlapping sets, whose union contains the support of μ. At the $(n+1)$st stage, each of these sets are subdivided and so on. For example, the support of the Cantor measure is contained in $[0,1]$, say $\mathcal{K}_0$. The collection of sets at the first level is

$$\mathcal{K}_1 = \left[0, \tfrac{1}{3}\right] \cup \left[\tfrac{2}{3}, 1\right],$$

and the second level is

$$\mathcal{K}_2 = \left[0, \tfrac{1}{9}\right] \cup \left[\tfrac{2}{9}, \tfrac{1}{3}\right] \cup \left[\tfrac{2}{3}, \tfrac{7}{9}\right] \cup \left[\tfrac{8}{9}, 1\right],$$

and so on. The Cantor set is $\mathcal{K} = \bigcap_{n=0}^{\infty} \mathcal{K}_n$ and is the support of the Cantor measure. We need notation to describe this process in a more general setting.

A similar framework can be used as that for the multinomial measures in Chapter 3. Let $\Omega = \{0, 1, \cdots, b-1\}$, where b is a fixed integer ≥ 2, and let Ω^n be the set of all sequences of length n containing digits in Ω. Then we define a one to one relationship between elements of Ω^n and a collection of *non-overlapping* subsets (see Definitions A.1.1) of $\mathcal{X}$ by a mapping $J_n(\omega)$ such that $\mathcal{K}_n = \bigcup_{\omega \in \Omega^n} J_n(\omega)$; i.e., if $\omega, \omega' \in \Omega^n$ and $\omega \neq \omega'$, then $J_n(\omega)$ and $J_n(\omega')$ are non-overlapping. Thus $\mathcal{K}_n$ consists of b^n non-overlapping subsets that cover the support of μ.

Let Ω^∞ denote the set of all infinite sequences (one sided to the right) containing digits in Ω. We impose further structure by defining a projection of Ω^∞ onto Ω^n that extracts the first n digits from $\omega \in \Omega^\infty$, denoted by $\omega|n$. The subdivision is required to satisfy the further condition that, for all n and all $\omega \in \Omega^\infty$, $J_{n+1}(\omega|(n+1)) \subset J_n(\omega|n)$. Our aim is to define a measure μ with support $\mathcal{K} = \bigcap_{n=0}^{\infty} \mathcal{K}_n$. $\mathcal{K}_n$ will be a covering of the support of μ, and $\mathcal{K}_{n+1}$ will be a refined covering.

Define the *coding map* X as

$$X : \Omega^\infty \longrightarrow \mathcal{X},$$

in particular

$$X(\omega) = \bigcap_{n=1}^{\infty} J_n(\omega|n). \tag{6.1}$$

So far we have simply used the space of infinite sequences Ω^∞ like an index set to describe a nested covering system closing down onto the support of μ.

Now let $\mathcal{S}_n$ be the sub-σ-field generated by all subsets $J_n(\omega)$ of $\mathcal{K}_n$. Assuming that we have a method to define a measure μ_n for all $A \in \mathcal{S}_n$, then we have a measure space $(\mathcal{X}, \mathcal{S}_n, \mu_n)$. We will then define μ to be the inductive limit of μ_n.

One must be careful if one is to also think of Ω^∞ as part of a measure space. Recall that there is a one to one relationship between elements of Ω^n and the subsets of $\mathcal{K}_n$, and hence the sub-σ-field generated by all subsets of $\mathcal{K}_n$, denoted by $\mathcal{S}_n$, has a one to one relationship with that generated by the elements in Ω^n, called $\mathcal{F}_n$ say. Further, given that μ_n is defined, let $\rho_n(E) = \mu_n(J_n(E))$ where $E \in \mathcal{F}_n$. Hence the two measure spaces $(\Omega^\infty, \mathcal{F}_n, \rho_n)$ and $(\mathcal{X}, \mathcal{S}_n, \mu_n)$ are isomorphic, one being based on sequences, and the other being based on d-dimensional Euclidean space. Our following discussions use that one which is the most convenient in the given situation.

Now we want to impose more explicit structure on the size of the subsets $J_n(\omega)$ and the manner in which weight (not necessarily probability) is allocated to such subsets. Consider the mappings

$$W_i : \Omega^i \longrightarrow (0, \infty) \tag{6.2}$$

and

$$T_i : \Omega^i \longrightarrow (0, 1), \tag{6.3}$$

where $i = 1, \cdots, n$. W_i describes the allocation of weight or mass to the subsets $J_n(\omega)$, and T_i determines their size. Let μ_n be the mass distribution at the nth stage, and assume that the mass allocation and size of the sets can be described as

$$\mu_n[J_n(\omega)] = \prod_{i=1}^{n} W_i(\omega|i) \quad \text{and} \quad |J_n(\omega)| = \prod_{i=1}^{n} T_i(\omega|i), \tag{6.4}$$

respectively. Then let μ be the limiting mass allocation as $n \to \infty$.

To describe the multifractal behaviour of the limit measure μ, we observe the local growth or behaviour. In Chapters 4 and 5, we considered a mapping of the form $Y_n : \Omega^\infty \to \mathbb{R}$, where

$$Y_n(\omega) = \frac{\log \mu[J_n(\omega|n)]}{\log |J_n(\omega|n)|}, \tag{6.5}$$

where $|J_n(\omega|n)| \to 0$ as $n \to \infty$. Note that Y_n is different from that in Chapter 4, because in this chapter, the denominator $\log |J_n(\omega)|$ is not assumed to be constant for all subsets of $\mathcal{K}_n$.

In the cascade literature, a distinction is made between *random* cascades and *deterministic* cascades. Deterministic is meant in the sense that the mappings W_i and T_i are both explicitly defined, as was done with the multinomial measures in Chapter 3. Random cascades occur when one or both of W_i and T_i are random variables. One could have a 2×2 situation, where one can either subdivide according to a deterministic or random process in the above sense, and one could also allocate weight according to a random or deterministic process.

6.1.1 Definition

Assume that there are given mappings of the form in Equations 6.2 and 6.3, and that the measure μ_n on the subsets $J_n(\omega)$ and the size of the subsets $J_n(\omega)$ are given by Equation 6.4, for all $\omega \in \Omega^n$, and for $n = 1, 2, \cdots$. Also assume that the inductive limit of the measure μ_n exists, and is denoted by μ. Then the measure μ will be said to represent a *multiplicative cascade process*.

The local behaviour of μ_n, denoted by $Y_n^\ddagger(\omega)$, can be represented for all $\omega \in \Omega^\infty$ as

$$Y_n^\ddagger(\omega) = \frac{\log \mu_n[J_n(\omega|n)]}{\log |J_n(\omega|n)|} = \frac{\sum_{i=1}^{n} \log W_i(\omega|i)}{\sum_{i=1}^{n} \log T_i(\omega|i)}. \tag{6.6}$$

We will call it an *elementary multiplicative cascade process* if $|J_n(\omega)| = \delta_n$, where δ_n is constant for a given n, and a decreasing sequence as $n \to \infty$. $\quad\square$

Note the distinction between $Y_n^{\ddagger}(\omega)$ and $Y_n(\omega)$, where $Y_n^{\ddagger}(\omega)$ is the local behaviour of μ_n, and $Y_n(\omega)$ is the local behaviour of μ. In the case where the mappings W_i and T_i are deterministic, as described above, then $Y_n^{\ddagger}(\omega) = Y_n(\omega)$. In the case where the allocated weights (or subdividing) are random, as described above, then $\mu_n[J_n(\omega)] \neq \mu[J_n(\omega)]$ and hence $Y_n^{\ddagger}(\omega) \neq Y_n(\omega)$. In the case of the multinomial measures (Chapter 3), the mapping T_i was trivial, i.e., $T_i(\omega) = b^{-1}$ giving $|J_n(\omega)| = b^{-n}$. Further, the form of the mapping W_i was also explicitly defined, and hence $Y_n^{\ddagger}(\omega) = Y_n(\omega)$.

In Chapter 4, global averages, Rényi dimensions, the multifractal spectrum, and the partition sets were denoted by $\widetilde{\theta}(q)$, $\widetilde{D}_q$, $\widetilde{f}(y)$ and $\widetilde{F}(y)$, respectively. Similarly, in Chapter 5, they were denoted by $\theta(q)$, D_q, $f(y)$ and $F(y)$, respectively. In both Chapters 4 and 5 the width of all covers were the same. In this chapter, functions describing similar concepts will be defined, but will be denoted by $\theta^{\star}(q)$, $D_q^{\star}$, $f^{\star}(y)$ and $F^{\star}(y)$, respectively. In this chapter, we also want to consider random measures. This requires a more general definition for $\theta^{\star}(q)$ than that of $\widetilde{\theta}(q)$. For example, if one constructs a random measure μ on the b-adic intervals, then $\widetilde{\theta}(q)$ is random in nature. We want $\theta^{\star}(q)$ to be similar in nature to a cumulant generating function.

6.1.2 Definition - Global Averaging

For given values of n and q, $\theta_n^{\star}(q)$ is defined to be the solution to the equation

$$\mathrm{E}\left[\frac{\mu_n^q[J_n(\omega)]}{|J_n(\omega)|^{\theta_n^{\star}(q)}}\right] = b^{-n},$$

assuming that the expectation exists which is taken over all subsets of the nth level with equal probability b^{-n}. Equivalently, in a format like a cumulant generating function,

$$\mathrm{E}\left[\exp\left(q\sum_{i=1}^{n}\log W_i(\omega|i) - \theta_n^{\star}(q)\sum_{i=1}^{n}\log T_i(\omega|i)\right)\right] = b^{-n}.$$

Then $\theta^{\star}(q) = \lim_{n\to\infty}\theta_n^{\star}(q)$ assuming that this limit exists. The *multiplicative cascade Rényi dimensions* are defined for $q \neq 1$ as

$$D_q^{\star} = \frac{\theta^{\star}(q)}{q-1}.$$

□

Note the difference between $\widetilde{\theta}(q)$ and $\theta^{\star}(q)$, even when $|J_n(\omega)| = \log\delta_n$ for all ω. When μ is a random measure, then $\widetilde{\theta}(q)$ is random in nature (sometimes referred to as a *sample average*), whereas $\theta^{\star}(q)$ is not (sometimes referred to as an *ensemble average*). This will be discussed further in §6.3. However, the interpretation of $\theta^{\star}(q)$ is similar to both $\widetilde{\theta}(q)$ and $\theta(q)$, see Example 3.3.2.

When the partitioning at each step is regular as in Chapters 4 and 5, both $\theta(q)$

and $\widetilde{\theta}(q)$ have an important probabilistic interpretation. They were related to the rescaled cumulant generating function, and as such one can appeal to the theory of large deviations to describe the local behaviour, and justify Legendre transform relationships between the global averaging and local behaviour. In Chapter 4, under satisfactory conditions, the local behaviour of μ was described as

$$\lim_{n \to \infty} \frac{\log \Pr\{Y_n \in A\}}{\log \delta_n} = - \inf_{y \in A} I(y),$$

where δ_n was the size of the subsets at the nth stage and $I(y)$ was related to $\widetilde{\theta}(q)$ via a Legendre transform. Further, the multifractal spectrum, $\widetilde{f}(y)$, was defined in such a way that it was explicitly related to $I(y)$ (see Equation 4.7). However, in the present context, the subset sizes at each step are not a constant δ_n, and hence the results are not applicable. In the cascade context, we require a different definition of $f^\star(y)$.

6.1.3 Definition - Multifractal Spectrum

The multifractal spectrum of a multiplicative cascade process, denoted by $f^\star(y)$, is defined for $y \in \mathbb{R}$ as

$$f^\star(y) = \inf_q \left\{ qy - \theta^\star(q) \right\}.$$

□

In a sense, $f^\star(y)$ is the analogue of $\widetilde{f}(y)$ given by Equation 2.7. This is not strictly correct though. Recall that in Chapter 4, $C_0(q)$ was the rescaled cumulant generating function with the uniform sampling measure. Apart from constants and sign changes, it is related to $\widetilde{\theta}(q)$ given by Equation 4.6. However, $I_0(y)$ was *defined* as the Legendre transform of $C_0(q)$, and hence $I_0(y)$ (apart from constants and sign changes) is really the analogue of $f^\star(y)$. Note that $\widetilde{f}(y)$ in Chapter 4 was *defined* by Equation 2.7, and satisfied the Legendre transform relationships if the conditions of Corollary 4.3.5 were satisfied. In the present formulations, the definition of $\widetilde{f}(y)$ given by Equation 2.7, is not relevant, as $\log \delta_n$ may be different for different subsets of $\mathcal{K}_n$.

The remainder of this chapter is mainly given to discussing Moran and random cascade processes. Moran cascade processes are described in §6.2. In that case the construction of the support and measure allocation is performed in a deterministic manner as described above, though is more general than the multinomial measures in Chapter 3. In §6.3 results relating to random cascades are reviewed. The particular case studied here is where the measure allocation is random, but the division of subsets is the same as in Chapter 3. Consequently, the Gärtner-Ellis Theorem is still applicable, though one has the complication that $Y_n^\ddagger(\omega) \neq Y_n(\omega)$. Other cascade processes are briefly discussed in §6.4.

Much of the material in this chapter has been drawn from Cawley & Mauldin (1992), Holley & Waymire (1992) and Gupta & Waymire (1990, 1993).

6.2 Moran Cascade Processes

Many of the cascade processes in the literature stem from a result due to Moran (1946). He showed that if a set was repeatedly subdivided in a certain way, the Hausdorff dimension of the limiting set satisfied a particularly simple relationship. Sets that conform to such behaviour are self-similar. These ideas were further developed by Hutchinson (1981). Definitions of self-similar sets and their Hausdorff dimension can be found in Appendix A, §A.1.

The construction of a Moran fractal set is very similar to that of the Cantor set, in that scaled copies are made of the set at each division. The difference is that the scaling ratios are not required to be constant, and the initial or seed set can take a more general form than the unit interval. Moran cascade processes have been discussed by Cawley & Mauldin (1992) and Edgar & Mauldin (1992).

6.2.1 Definition - Moran Fractal Set (Moran, 1946; Hutchinson, 1981)

The symbol $\bowtie$ is to be interpreted as concatenation, i.e., if $\omega = (\omega_1, \cdots, \omega_n) \in \Omega^n$, then $\omega \bowtie k$ is to be interpreted as $(\omega_1, \cdots, \omega_n, k) \in \Omega^{n+1}$. Associated with each element $\omega \in \Omega = \{0, 1, \cdots, b-1\}$ is a similarity ratio, t_ω, where $0 < t_\omega < 1$. Let $\mathcal{K}_0 \subset \mathbb{R}^d$ be a *seed* set. It is assumed to be regular, i.e., $\mathcal{K}_0 = \text{closure}(\text{int}\mathcal{K}_0)$. The set $J_n(\omega)$, $\omega \in \Omega^n$, is determined recursively, as follows. Without loss of generality, let $|\mathcal{K}_0| = 1$.

1. $J_1(0), \cdots, J_1(b-1)$ are each *non-overlapping* similarities of $\mathcal{K}_0$, with similarity ratios $t_0, \cdots, t_{b-1}$, respectively. (Non-overlapping is to be interpreted as satisfying the open set condition, Edgar & Mauldin, 1992, page 606; see Definition A.1.1.)

2. If $J_n(\omega)$ has been determined for $\omega \in \Omega^n$, then $J_{n+1}(\omega \bowtie 0), J_{n+1}(\omega \bowtie 1), \cdots, J_{n+1}(\omega \bowtie (b-1))$ are *non-overlapping* subsets of $J_n(\omega)$ such that for each $k \in \Omega$, $J_{n+1}(\omega \bowtie k)$ is geometrically similar to $J_n(\omega)$ via a similarity map with reduction ratio t_k.

For $n \geq 1$, $\mathcal{K}_n = \bigcup_{\omega \in \Omega^n} J_n(\omega)$. The *Moran Fractal Set* is then defined as

$$\mathcal{K} = \bigcap_{n=0}^{\infty} \mathcal{K}_n.$$

□

6.2.2 Theorem

The Hausdorff and box dimensions of the Moran fractal set are equal, moreover,

$$\dim_H \mathcal{K} = \dim_B \mathcal{K} = s,$$

where s is the solution to $\sum_{\omega \in \Omega} t_\omega^s = 1$.

Proof. Follows from Theorem A.1.5.

□

A Moran Fractal Set

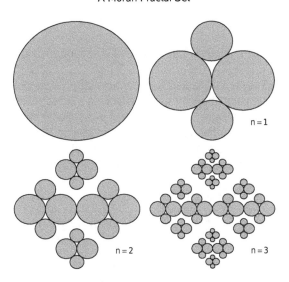

Figure 6.1 *An example of a Moran fractal set, with the seed set* $\mathcal{K}_0$ *in the top left. Also shown are* $\mathcal{K}_1$, $\mathcal{K}_2$ *and* $\mathcal{K}_3$. *The similarity ratios are* $t_0 = \frac{1}{3}, t_1 = \frac{1}{2}, t_2 = \frac{1}{3}$ *and* $t_3 = \frac{1}{2}$. *The Hausdorff dimension of* $\mathcal{K}$ *is* 1.605525 $\cdots$.

6.2.3 Example

Let the circle in the upper left of Figure 6.1 represent a seed set $\mathcal{K}_0$. Let $b = 4$, with the four similarities $J_1(0), \cdots, J_1(3)$ of $\mathcal{K}_1$ be the circles in the top right $(n = 1)$ of Figure 6.1, in the order top, right, bottom and left. These have similarities $t_0 = \frac{1}{3}, t_1 = \frac{1}{2}, t_2 = \frac{1}{3}$ and $t_3 = \frac{1}{2}$. $\mathcal{K}_2$ and $\mathcal{K}_3$ are represented in Figure 6.1 at the bottom left and bottom right respectively. From Theorem 6.2.2 the resulting Moran fractal set $\mathcal{K} = \bigcap_{n=0}^{\infty} \mathcal{K}_n$ has Hausdorff dimension 1.605525 $\cdots$. $\square$

6.2.4 Moran Cascade Measure

Associated with each $\omega \in \Omega$ is a non-zero probability p_ω such that $\sum_{\omega \in \Omega} p_\omega = 1$, and a similarity ratio $0 < t_\omega < 1$. Then the mappings W_i and T_i, given by Equations 6.2 and 6.3 respectively, are defined explicitly as

$$W_i(\omega) = p_{\omega_i} \quad \text{and} \quad T_i(\omega) = t_{\omega_i}.$$

Thus

$$\mu_n[J_n(\omega)] = \prod_{i=1}^{n} W_i(\omega|i) = p_{\omega_1} p_{\omega_2} \cdots p_{\omega_n},$$

and

$$|J_n(\omega)| = \prod_{i=1}^{n} T_i(\omega|i) = t_{\omega_1} t_{\omega_2} \cdots t_{\omega_n}.$$

The measure μ is the inductive limit of μ_n and is supported by the Moran fractal set $\mathcal{K}$.   □

Note the difference with Chapter 3, where here all p_ω's are non-zero. It could be made completely compatible by simply attaching a subscripted zero to the Ω's in this chapter, since in Chapter 3, $\Omega_0 = \{\omega : p_\omega > 0\}$. For simplicity of notation, we have not done this. Further, in the construction used in Chapter 3, zero values of p_ω were necessary to create gaps in the support of μ. This is not necessary here, as gaps can be created by having different similarity ratios, which need not necessarily sum to one.

It follows from Equation 6.6 that

$$Y_n^{\ddagger}(\omega) = \frac{\log \mu_n[J_n(\omega|n)]}{\log |J_n(\omega|n)|} = \frac{\sum_{i=1}^{n} \log p_{\omega_i}}{\sum_{i=1}^{n} \log t_{\omega_i}}.$$

In the case of the Moran cascades $\mu[J_n(\omega)] = \mu_n[J_n(\omega)]$ and so $Y_n(\omega) = Y_n^{\ddagger}(\omega)$. In Chapters 4 and 5, the space $\mathcal{X}$ was partitioned according to the local powerlaw behaviour of the measure μ, that is, $F(y) \subset \mathcal{X}$ and $\widetilde{F}(y) \subset \mathcal{X}$. Here we also want to define $F^\star(y)$ in such a way that $F^\star(y) \subset \mathcal{X}$.

6.2.5 Definition - Partition Set

The partition set $F^\star(y)$, for $y > 0$, is defined as $F^\star(y) = X(G(y))$, where

$$G(y) = \left\{ \omega \in \Omega^\infty : \lim_{n \to \infty} Y_n(\omega) = y \right\},$$

and X is the coding map in Equation 6.1.   □

6.2.6 Note

Recall that in Chapter 4, $B_{\delta_n}(x)$ was defined to be half open. Thus for each $x \in \mathcal{X}$, $B_{\delta_n}(x)$ is unique, hence the set

$$\widetilde{F}(y) = \left\{ x \in \mathcal{X} : \lim_{n \to \infty} \frac{\log \mu[B_{\delta_n}(x)]}{\log \delta_n} = y \right\}$$

is properly defined, and the local behaviour can be determined.

The Moran fractal set $\mathcal{K}$ is constructed as the intersection of nested closed sets with positive μ-measure as $n \to \infty$. Hence $\mathcal{K}$ is the support of μ and $\mathcal{K} \subseteq \mathcal{X}$. Further note that $J_n(\omega)$ is closed because $\mathcal{K}_0$ is closed. However, the definition using the coding map (Equation 6.1) ensures that $F^\star(y)$ is unambiguously defined.   □

The function $\theta^\star(q)$ for the Moran cascade process satisfies a particularly simple relationship. Cawley & Mauldin (1992) refer to $\theta^\star(q)$ as an 'auxiliary' function.

However, as already shown, it has a similar interpretation to the functions $\tilde{\theta}(q)$ and $\theta(q)$ in Equations 2.2 and 2.5, respectively.

6.2.7 Theorem

In the case of the Moran cascade processes, $\theta^\star(q)$, where $q \in \mathbb{R}$, is the unique solution to

$$\sum_{k=0}^{b-1} p_k^q t_k^{-\theta^\star(q)} = 1.$$

Proof. From Definition 6.1.2,

$$\sum_{\omega \in \Omega^n} \mu_n^q[J_n(\omega)]\,|J_n(\omega)|^{-\theta_n^\star(q)} = \sum_{\omega \in \Omega^n} p_{\omega_1}^q \cdots p_{\omega_n}^q\, t_{\omega_1}^{-\theta_n^\star(q)} \cdots t_{\omega_n}^{-\theta_n^\star(q)}$$

$$= \left(\sum_{\omega \in \Omega} p_\omega^q t_\omega^{-\theta_n^\star(q)} \right)^n.$$

This is the sum over all b^n subsets of $\mathcal{K}_n$, hence should be equal to one. The result then follows. $\qquad\square$

When $q = 0$, it follows from Theorem 6.2.2 that $D_0^\star = -\theta^\star(0) = \dim_H \mathcal{K} = \dim_H \operatorname{supp}(\mu)$. In the case of multinomial measures, $t_i = b^{-1}$ and hence $\theta^\star(q) = \tilde{\theta}(q)$.

6.2.8 Theorem - Multifractal Spectrum

The multifractal spectrum, $f^\star(y)$ for $y \in \mathbb{R}$, of a Moran cascade process can be calculated as

$$f^\star(y_q^\star) = q y_q^\star - \theta^\star(q),$$

where $y_q^\star$ is the derivative of $\theta^\star(q)$, i.e.,

$$y_q^\star = \frac{d}{dq}\theta^\star(q) = \frac{\displaystyle\sum_{k=0}^{b-1} p_k^q t_k^{-\theta^\star(q)} \log p_k}{\displaystyle\sum_{k=0}^{b-1} p_k^q t_k^{-\theta^\star(q)} \log t_k}.$$

Proof. From Definition 6.1.3, $f^\star(y) = \inf_q \{qy - \theta^\star(q)\}$. We have two functions of q, one linear with slope y, the other being $\theta^\star(q)$, see Figure 3.4. For a given value of q, the infimum occurs when both have the same slope, i.e., when $y = y_q^\star$. $\qquad\square$

The function $y_q^\star$ is the analogue of y_m in Chapters 4 and 5, with $m = q$.

6.2.9 Theorem - Hausdorff Dimensions (Cawley & Mauldin, 1992, Theorem 2.1)

Let $y > 0$, then $\dim_H F^\star(y) = f^\star(y)$, where $F^\star(y)$ is the partition set as in Definition 6.2.5 and $f^\star(y)$ is the multifractal spectrum as in Theorem 6.2.8.

Proof. An outline of the proof follows. The proof is in two parts: $\dim_H F^\star(y) \leq f^\star(y)$ is shown by using a Vitali covering argument. Then $\dim_H F^\star(y) \geq f^\star(y)$ is shown by a geometric argument *that is intrinsic to this particular setup.* In the course of the proof, Cawley & Mauldin (1992) introduce *auxiliary measures,* denoted by $\Pi_m^{(\infty)}$ on the space $(\Omega^\infty, \mathcal{F}_n)$ defined by associating the probabilities $(p_0 t_1^{-\theta^\star(m)}, \cdots, p_{b-1} t_{b-1}^{-\theta^\star(m)})$ with each $\omega \in \Omega$. By the definition of $\theta^\star(m)$, these probabilities sum to one. Associated probabilities with each $\omega \in \Omega^n$ are calculated as the relevant products.

In Chapters 4 and 5 we required that $Y_n \to y_m$ $\nu_m^{(\infty)}$-a.s. where $\nu_m^{(\infty)}$ was the analogue of $\Pi_m^{(\infty)}$. Cawley & Mauldin (1992) also rely on a strong law of large numbers, achieved by appealing to Birkhoff's ergodic theorem (Walters, 1982, page 35) as follows. For $\Pi_m^{(\infty)}$ almost all $\omega \in \Omega^\infty$,

$$\lim_{n\to\infty} \frac{1}{n} \sum_{i=1}^{n} \log W_i(\omega|i) = -\sum_{k=0}^{b-1} p_k^m t_k^{-\theta^\star(m)} \log p_k.$$

Similarly, for $\Pi_m^{(\infty)}$ almost all $\omega \in \Omega^\infty$,

$$\lim_{n\to\infty} \frac{1}{n} \sum_{i=1}^{n} \log T_i(\omega|i) = -\sum_{k=0}^{b-1} p_k^m t_k^{-\theta^\star(m)} \log t_k.$$

Taking ratios, it then follows that $\lim_{n\to\infty} Y_n(\omega) = y_m^\star$ $\Pi_m^{(\infty)}$-a.s., consistent with Chapter 4. Therefore Ω^∞ can be partitioned as $\Pi_m^{(\infty)}[G(y_m)] = 1$. Hence the choice of auxiliary probabilities $(p_0 t_1^{-\theta^\star(m)}, \cdots, p_{b-1} t_{b-1}^{-\theta^\star(m)})$ produces an analogous argument to that of Chapters 4 and 5 where the sampling measure is used. □

A disadvantage of the definition of $F^\star(y)$ is that the determination of when $x \in F^\star(y)$ depends on knowing some sequence of sets $J_n(\omega)$ of construction closing down on x. For disjoint type Moran constructions, Cawley & Mauldin (1992) showed that the local behaviour of the measure μ can be expressed in terms of a sphere centred at x.

6.2.10 Theorem (Cawley & Mauldin, 1992)

Consider a Moran cascade process generated by pairwise *disjoint* mappings (i.e., $J_n(\omega) \cap J_n(\omega') = \emptyset$ for all $\omega \neq \omega' \in \Omega^n$, and for all n). Then $F^\star(y) = F(y)$. □

This is a general version of Theorems 5.5.1 and 5.5.2. Note also that the subsets of $\mathcal{K}_n$ must have gaps between them. This is because each $J_n(\omega)$ is a closed interval, and if the subsets are disjoint, then there must be a non-zero distance between any pair.

In a series of papers, Mandelbrot and co-authors discuss multiplicative cascade processes with anomalous behaviour. They refer to them as 'exactly self-similar left-sided' processes, where the multifractal spectrum is increasing over its entire range (see Mandelbrot, 1990b; Mandelbrot et al., 1990 and Riedi & Mandelbrot, 1995).

6.3 Random Cascades

This case has been discussed by Holley & Waymire (1992) and Gupta & Waymire (1990, 1993). In this section, the process starts with $\mathcal{K}_0 = [0, 1]$, which is then divided into $b \geq 2$ intervals of equal length. Each of these subintervals are further divided ad infinitum. The interval $J_n(\omega)$ is defined as

$$J_n(\omega) = \left[\sum_{j=1}^n \frac{\omega_j}{b^j}, \ \frac{1}{b^n} + \sum_{j=1}^n \frac{\omega_j}{b^j} \right].$$

In this case, the mapping T_i in Equation 6.3, is fixed and constant; i.e., for all i and $\omega \in \Omega^i$, $T_i(\omega) = b^{-1}$, and thus $J_n(\omega)$ represents a specific subinterval of length b^{-n}.

As in Chapter 4, it will sometimes be more convenient to use an arbitrary enumeration of the intervals, rather that $J_n(\omega)$. In this case, we will refer to them as $B_{\delta_n}(k)$ where $k = 1, \cdots, b^n$ and $\delta_n = b^{-n}$. The box to which a point $x \in [0, 1]$ belongs should logically be denoted as $B_{\delta_n}(k(x))$, but to avoid notation becoming too clumsy, it will be written as $B_{\delta_n}(x)$, when in the given context, it cannot be confused with $B_{\delta_n}(k)$. As in Chapter 4, $B_{\delta_n}(k)$, $k = 1, \cdots, b^n$, are half open to the right, whereas $J_n(\omega)$ are closed intervals. However, as with the multinomial measures, the resultant measure on the boundaries of the intervals will be zero, and hence this does not have an effect.

Weight is to be allocated to the subintervals randomly, and hence the resultant measure is not a probability measure. That is, the mapping $W_i(\omega)$ given by Equation 6.2 is a random variable.

6.3.1 Construction of Random Measures

A measure is constructed on the interval $[0, 1]$ as follows. Let

$$W_i(\omega) \equiv W(\omega_1, \cdots, \omega_i),$$

where $W_i(\omega)$, $i = 1, 2, \cdots$, are i.i.d. non-negative random variables with mean b^{-1}. Define the random measure μ_n with density (Radon-Nikodym derivative) as

$$\mu_n(dx) = b^n W_1(\omega) W_2(\omega) \cdots W_n(\omega) dx$$

where $n \geq 1$ and $x \in J_n(\omega)$. Thus

$$
\begin{aligned}
\mu_n[J_n(\omega)] &= \int_{J_n(\omega)} b^n W_1(\omega) W_2(\omega) \cdots W_n(\omega) dx \\
&= W_1(\omega) W_2(\omega) \cdots W_n(\omega).
\end{aligned}
$$

The inductive limit measure will be denoted by μ. $\qquad\qquad\square$

In what follows, the variable W is assumed to have the same distribution as $W_i(\omega)$. Note that Gupta & Waymire (1993) stipulate that $E[W] = 1$. We require that $E[W] = b^{-1}$, ensuring that these processes can be described as in Equation 6.6. Below we quote some results, mainly from Holley & Waymire (1992), changing where necessary for the effect of scaling W by b^{-1}.

6.3.2 Definition (Holley & Waymire, 1992, Definition 2.1)

The random variable W is said to be *strongly bounded below* if there is a positive number a such that $\Pr\{W > a\} = 1$. Similarly, W is said to be *strongly bounded above* if $\Pr\{W < 1\} = 1$. $\qquad\qquad\square$

6.3.3 Theorem (Kahane & Peyrière, 1976; Holley & Waymire, 1992)

Let W denote a random variable with the same distribution as the $W_i(\omega)$'s, and let $Z_\infty = \mu([0,1])$.

1. If $E[W \log_b W] < 0$, then $E[Z_\infty] > 0$, and conversely. The condition that $E[Z_\infty] > 0$ is equivalent to $E[Z_\infty] = 1$.

2. Let $q > 1$. Then Z_∞ has a finite moment of order q iff $q < q_{\text{crit}} = \sup\{q \geq 1 : -\log_b E[W^q] > 1\}$. Moreover, $E[Z_\infty^q] < \infty$ for all $q > 0$ iff W is essentially bounded by 1 (i.e., $\|W\|_\infty < 1$) and $\Pr\{W = 1\} < 1/b$.

3. Assume that $E[Z_\infty \log_b Z_\infty] < \infty$. Then μ is a.s. supported by the random set

$$
\text{supp}(\mu) = \left\{ x \in [0,1] : \lim_{n \to \infty} \frac{\log \mu[B_{\delta_n}(x)]}{\log \delta_n} = -b\, E[W \log_b W] \right\}
$$

with Hausdorff dimension $-b\, E[W \log_b W]$. $\qquad\qquad\square$

6.3.4 Corollary (Holley & Waymire, 1992, Corollary 2.5)

If the cascaded variable W is strongly bounded above, then Z_∞ has all moments of positive order. If the cascaded variable W is strongly bounded below, then Z_∞ has negative moments of all orders. $\qquad\qquad\square$

Since the allocation of mass is random, then the mass at the nth generation in a particular interval, $\mu_n[J_n(\omega)]$, will not be the same as the limiting mass allocation $\mu[J_n(\omega)]$. In the random cascade literature, μ_n is often referred to as a *bare cascade measure*, while μ is referred to as a *dressed cascade measure*. The distribution of the two measures are related as stated in the following proposition.

6.3.5 Proposition (Holley & Waymire, 1992, Proposition 2.3)

Let $B_{\delta_n}(k)$, $k = 1, \cdots, b^n$, be an arbitrary enumeration of the nth generation b-adic intervals $J_n(\omega)$ of $[0,1]$, and $\overset{d}{=}$ denote equality of probability distributions. Then the following hold.

1.
$$\mu_m[B_{\delta_n}(k)] \overset{d}{=} Z^{(n)}_{m-n}(k)\mu_n[B_{\delta_n}(k)],$$

   where $m \geq n$, $k = 1, \cdots, b^n$; and $Z^{(n)}_{m-n}(k)$ has the same distribution as the total mass $\mu_{m-n}([0,1])$ and is independent of $\mu_n[B_{\delta_n}(k)]$.

2.
$$\mu[B_{\delta_n}(k)] \overset{d}{=} Z^{(n)}_{\infty}(k)\mu_n[B_{\delta_n}(k)],$$

   where $m \geq n$, $k = 1, \cdots, b^n$; and $Z^{(n)}_{\infty}(k)$ has the same distribution as the total mass $\mu([0,1])$ and is independent of $\mu_n[B_{\delta_n}(k)]$. $\square$

Holley & Waymire (1992) characterise the process by the function $\chi_b(q) = \log_b \mathrm{E}[W^q] + 1$. In the literature $\chi_b(q)$ is referred to as the MKP (Mandelbrot-Kahane-Peyrière) function. It turns out that $\chi_b(q)$ is just the appropriate form of $-\theta^\star(q)$, as in Definition 6.1.2, for a random cascade process.

6.3.6 Theorem

If μ is the inductive limit of a random cascade process, then $\theta^\star(q)$ as in Definition 6.1.2 is

$$\theta^\star(q) = -\log_b \mathrm{E}[W^q] - 1.$$

Proof. In the present situation $\mu_n[J_n(\omega)] = W_1(\omega)\cdots W_n(\omega)$, where the W_i's are i.i.d. random variables and $|J_n(\omega)| = b^{-n}$. Hence

$$\begin{aligned}
b^{-n} &= \mathrm{E}\left[\mu_n^q[J_n(\omega)]\,|J_n(\omega)|^{-\theta_n^\star(q)}\right] \\
&= \left(b^{-n}\right)^{-\theta_n^\star(q)} \mathrm{E}[W_1^q(\omega)\cdots W_n^q(\omega)] \\
&= \left(b^{-n}\right)^{-\theta_n^\star(q)} \mathrm{E}[W^q]^n
\end{aligned}$$

where W has the same distribution as the W_i's. Rearranging, the result follows.
                                                                              $\square$

6.3.7 Example - Log-Lévy Generators

A class of cascade generators is given by $W = \exp(-Z)$ where Z is a Lévy stable random variable. Lovejoy & Schertzer have co-authored number of papers (e.g., see Schertzer & Lovejoy, 1987, 1989; and Lovejoy & Schertzer, 1985, 1990) where they advocate processes of this form to describe rain fields. Gupta &

Waymire (1993, Equation 3.15) give $E[W^q]$ for these processes, as

$$\log E[W^q] = \begin{cases} -c\sigma^\alpha q^\alpha + \xi q & \text{if } 0 < \alpha < 1 \\ cq \log q - \xi q & \text{if } \alpha = 1 \\ c\sigma^\alpha q^\alpha - \xi q & \text{if } 1 < \alpha \leq 2, \end{cases}$$

where α is the characteristic exponent, $q > 0$ when $0 < \alpha < 2$, $\xi \in \mathbb{R}$ is a location parameter, $\sigma > 0$ is a scale parameter, and c is a positive constant.

The case where $0 < \alpha < 1$ follows from Feller (1971, §XIII.6, Theorem 1), because $E[W^q]$ is the Laplace transform of the probability density of Z. The extension to other cases was done by Schertzer & Lovejoy (1987, Appendix C). These generators are often described as being 'universal' (e.g., Lovejoy & Schertzer, 1990), though as pointed out by Gupta & Waymire (1990, page 259), all multiplicative generators do not lie in the domain of attraction of a stable law. Also note that this class of generators does not satisfy the strong boundedness conditions of Definition 6.3.2.   □

Random Cascades: Large Deviation Formalism

We have already noted the difference between the bare and dressed (limiting) measures, and that $\mu_n[J_n(\omega)] \neq \mu[J_n(\omega)]$. Thus there is a choice as to whether the large deviation theorems are applied to μ_n or μ. The process we have used in Chapters 4 and 5 starts by evaluating the rescaled cumulant generating function $C(q)$, and deducing from that the entropy function $I(y)$ via a Legendre transform. Given that μ_n is explicitly defined in terms of sums of random variables $\log W_i$, it is easiest to evaluate the global and local behaviour of μ_n, then subsequently relate its local behaviour to that of μ.

Appealing to the notation of Appendix B, let $\delta_n = b^{-n}$ and $a_n = -\log \delta_n = n \log b$, where $n = 1, 2, \cdots$. Also define the mapping

$$U_n : \Omega^\infty \longrightarrow \mathbb{R},$$

where

$$U_n(\omega) = -\log \mu_n[J_n(\omega|n)] = -\sum_{i=1}^n \log W_i(\omega|i).$$

The measure μ_n is random, and constructed in the manner described in §6.3.1. Further, define $Y_n^\ddagger(\omega)$ as

$$Y_n^\ddagger(\omega) = \frac{U_n(\omega)}{a_n} = \frac{\log \mu_n[J_n(\omega|n)]}{\log \delta_n} = \frac{-1}{n} \sum_{i=1}^n \log_b W_i(\omega|i).$$

6.3.8 Rescaled Cumulant Generating Function

As in Chapter 4, consider the case where each b-adic interval is sampled with equal weight. We denote the corresponding expectation by E_0. The unsubscripted expectation below is taken with respect to the probability distribution of W. From Equation B.5,

$$
\begin{aligned}
C_0(q) &= \lim_{n\to\infty} \frac{1}{a_n} \log \mathrm{E}_0\left[e^{qU_n(\omega)}\right] \\
&= \lim_{n\to\infty} \frac{1}{n} \log_b \mathrm{E}_0\left[\mu_n^{-q}[J_n(\omega)]\right] \\
&= \lim_{n\to\infty} \frac{1}{n} \log_b \mathrm{E}\left[W_1^{-q}\cdots W_n^{-q}\right] \\
&= \log_b \mathrm{E}\left[W^{-q}\right].
\end{aligned}
$$

Note that $C_0(q) = -\theta^\star(-q) - 1$. The derivative of $\theta^\star(q)$, denoted by $y_q^\star$, is given by

$$
y_q^\star = \frac{d}{dq}\theta^\star(q) = \frac{-\mathrm{E}[W^q \log_b W]}{\mathrm{E}[W^q]}. \tag{6.7}
$$

Using Equation B.6, the entropy function is defined as

$$
I_0(y) = \sup_q\{qy - C_0(q)\} = 1 - f^\star(y).
$$

$\square$

6.3.9 Theorem - Large Deviation Bounds

If $C_0(q)$ satisfies the Extended Hypotheses 4.2.2, then

$$
\lim_{\epsilon\to 0}\lim_{n\to\infty} \frac{1}{n} \log_b \Pr\left\{-\epsilon < Y_n^\ddagger - y \le \epsilon\right\} = -I_0(y) \qquad \text{for } y \in \mathrm{int}\mathcal{D}(I).
$$

Further, $C_0(q) = \sup_y\{qy - I_0(y)\}$.

Proof. The powerlaw decay follows in the same manner as in Theorem 4.2.4. The Legendre transform follows from Theorem B.3.17. $\square$

The above theorem describes the local behaviour of the *bare* random measure. We also require the local behaviour of the dressed random measure μ.

6.3.10 Corollary

If $C_0(q)$ satisfies the Extended Hypotheses 4.2.2 and $\mathrm{E}[W \log_b W] < 0$, then

$$
\lim_{\epsilon\to 0}\lim_{n\to\infty} \frac{1}{n} \log_b \Pr\left\{-\epsilon < Y_n - y \le \epsilon\right\} = -I_0(y) \qquad \text{for } y \in \mathrm{int}\mathcal{D}(I).
$$

Proof. It follows from Proposition 6.3.5 that

$$
\Pr\left\{-\epsilon < Y_n - y \le \epsilon\right\} = \Pr\left\{y - \epsilon < \frac{\log Z_\infty^{(n)}(k)}{\log \delta_n} + Y_n^\ddagger \le y + \epsilon\right\}.
$$

Given $\mathrm{E}[W \log_b W] < 0$, it follows from Theorem 6.3.3 that the measure μ is

non-degenerate and $E[Z_\infty] = 1$. The result then follows from Theorem 6.3.9.
$\square$

Note that Gupta & Waymire (1993, §3b) derive the same result directly using Chernoff's theorem of large deviations.

Random Cascades: Ensemble and Sample Averages

The function $\widetilde{\theta}(q)$, defined by Equation 2.2, describes the powerlaw behaviour of averages of the form $\sum_k' \mu^q[B_{\delta_n}(k)]$ as $n \to \infty$. Since μ is a random measure, then $\widetilde{\theta}(q)$ is also random in nature and will not be functionally related to $\theta^\star(q)$ in a non-stochastic manner. We consider the limiting behaviour of $\widetilde{\theta}(q)$ below. In the literature on random cascades, $\widetilde{\theta}(q)$ is referred to as a *spatial or sample average*, while $C_0(q)$ and $\theta^\star(q)$ are referred to as *ensemble averages*.

Note that we have used $+ \log \delta_n$ in the denominator of the definition of $\widetilde{\theta}(q)$. In the literature on random cascades it tends to be $- \log \delta_n$, whereas in non-random cascades $+ \log \delta_n$. We have required our definition to be consistent throughout the book.

6.3.11 Theorem (Holley & Waymire, 1992, Theorem 2.7)

Assume that W is strongly bounded above and below and $E[W^{2q}]/E[W^q]^2 < b$ for a given q. Then $\widetilde{\theta}(q) = \theta^\star(q)$ with probability 1.
$\square$

6.3.12 Theorem (Holley & Waymire, 1992, Theorem 2.8)

Assume that W is strongly bounded above and below and $E[W^{2q}]/E[W^q]^2 < b$ for all q. If the multifractal spectrum $\widetilde{f}(y)$ exists, then $f^\star(y)$ is the *closed convex hull* of $\widetilde{f}(y)$. Further,

$$\theta^\star(q) = \inf_y \left\{ qy - \widetilde{f}(y) \right\}.$$

$\square$

Recall that the Legendre transform relationships between $C_0(q)$ and $I_0(y)$ work in both directions. Note also that the log-Lévy generators of §6.3.7 do not satisfy the strong boundedness conditions, and hence the above results may not necessarily hold.

Random Cascades: Hausdorff and Box Dimensions

Here the method that Holley & Waymire (1992) used to determine the Hausdorff dimensions is outlined. The method of Chapter 4 is quite similar, though uses a family of sampling measures. In the random cascade context, the measure μ will not be a probability measure, and hence Holley & Waymire (1992) construct a family of dual cascade processes.

6.3.13 Lemma (Holley & Waymire, 1992)

Assume that the cascaded random variable W is strongly bounded above and below and has mean b^{-1}. Consider those values of y such that $\inf_q \{qy - \theta^\star(q)\} = f^\star(y) > 0$. For each y, $y_q^\star = y$ (Equation 6.7) has a unique solution at $q = \beta(y)$ where $\beta(y)$ is a function of y. Moreover,

$$f^\star(y) = \inf_q \{qy - \theta^\star(q)\} = \beta(y)y - \theta^\star(\beta(y)). \tag{6.8}$$

Proof. Follows a similar argument as in Theorem 6.2.8. ☐

6.3.14 Theorem (Holley & Waymire, 1992, Theorem 2.6)

Assume that the cascaded random variable W is strongly bounded above and below and has mean b^{-1}. Also assume that

$$E\left[\left(\frac{W}{\|W\|_\infty}\right)^{\beta(y)}\right] > \frac{1}{b},$$

where $\|W\|_\infty$ denotes the essential supremum of W and $\beta(y)$ is determined by Equation 6.8 (hence, each y satisfies $f^\star(y) > 0$). Then $\dim_H \widetilde{F}(y) = \inf\{qy - \theta^\star(q)\} = f^\star(y)$, where

$$\widetilde{F}(y) = \left\{x \in [0, 1] : \lim_{n \to \infty} \frac{\log \mu[B_{\delta_n}(x)]}{\log \delta_n} = y\right\}.$$

Proof. An outline of the proof is as follows. W is a strongly bounded cascade variable with mean one. For each y such that $\inf_q \{qy - \theta^\star(q)\} = f^\star(y) > 0$, construct a dual cascaded variable W_β, distributed as

$$W_\beta \overset{d}{=} \frac{W^\beta}{b^{\beta+1}E[W^\beta]},$$

where $y_\beta^\star = y$ (Equation 6.7). That is, we replace values of $W(\omega_1, \cdots, \omega_n)$ sample point by sample point with $W_\beta(\omega_1, \cdots, \omega_n)$ for all $(\omega_1, \cdots, \omega_n) \in \Omega^n$. Let $\mu_{\infty,\beta}$ be the resultant cascade and let $Z_{\infty,\beta}$ denote the total mass. The proof can then be divided into three parts.

1. Existence and non-triviality of the cascaded measure $\mu_{\infty,\beta}$:
   $E[Z_{\infty,\beta}^q] < \infty$ for all $q > 1$, $\mu_{\infty,\beta}$ is non-trivial and $\mu_{\infty,\beta}([0, 1])$ is positive with probability 1.

2. Partitioning of the space.

   (a) It can be shown that, with probability 1

$$\sup_{1 \leq i \leq b^n} \frac{\log Z_\infty^{(n)}(i)}{n} \to 0 \qquad \text{as } n \to \infty,$$

where $Z_\infty^{(n)}(i)$, for $1 \leq i \leq b^n$, are i.i.d. distributed as Z_∞; and

$$\sup_{1 \leq i \leq b^n} \frac{\log Z_{\infty,\beta}^{(n)}(i)}{n} \to 0 \quad \text{as } n \to \infty,$$

where $Z_{\infty,\beta}^{(n)}(i)$, for $1 \leq i \leq b^n$, are i.i.d. distributed as $Z_{\infty,\beta}$.

(b) Let D_1 and D_β be the sets containing events where the above limits fail respectively. Since $D_1 \cup D_\beta$ has probability zero, one simply considers $[0,1] \setminus (D_1 \cup D_\beta)$.

(c) By definition, $x \in \widetilde{F}(y)$ iff

$$\lim_{n \to \infty} \frac{\log \mu[B_{\delta_n}(x)]}{\log \delta_n} = y. \tag{6.9}$$

From Proposition 6.3.5, $\mu[B_{\delta_n}(x)] \overset{d}{=} Z_\infty \mu_n[B_{\delta_n}(x)]$. Therefore, for $x \in \widetilde{F}(y) \cap D_1^c \cap D_\beta^c$, where D_1^c is the complement of D_1,

$$\lim_{n \to \infty} \frac{\log \mu[B_{\delta_n}(x)]}{\log(b^{-n})} \overset{d}{=} \lim_{n \to \infty} \frac{\log Z_\infty + \log \mu_n[B_{\delta_n}(x)]}{\log(b^{-n})}$$

$$\overset{d}{=} \lim_{n \to \infty} \frac{-1}{n}\left(\log_b Z_\infty - n + \sum_{i=1}^{n} \log_b W_i\right)$$

$$\overset{d}{=} 1 - \lim_{n \to \infty} \frac{1}{n}\sum_{i=1}^{n} \log_b W_i. \tag{6.10}$$

(d) For those $x \in \widetilde{F}(y)$, we also evaluate the local behaviour of the measure $\mu_{\infty,\beta}$. From Proposition 6.3.5, $\mu_{\infty,\beta}[B_{\delta_n}(x)] \overset{d}{=} Z_{\infty,\beta}\mu_{n,\beta}[B_{\delta_n}(x)]$. Therefore, for $x \in \widetilde{F}(y) \cap D_1^c \cap D_\beta^c$,

$$\lim_{n \to \infty} \frac{\log \mu_{\infty,\beta}[B_{\delta_n}(x)]}{\log(b^{-n})}$$

$$\overset{d}{=} \lim_{n \to \infty} \frac{\log Z_{\infty,\beta} + \log \mu_{n,\beta}[B_{\delta_n}(x)]}{\log(b^{-n})}$$

$$\overset{d}{=} \lim_{n \to \infty} \left(\beta + 1 - \frac{\log_b Z_{\infty,\beta}}{n} + \log_b \mathrm{E}[W^\beta] - \frac{1}{n}\sum_{i=1}^{n} \log_b W_i^\beta\right)$$

$$\overset{d}{=} \beta\left(1 - \lim_{n \to \infty} \frac{1}{n}\sum_{i=1}^{n} \log_b W_i\right) + \log_b \mathrm{E}[W^\beta] + 1.$$

It then follows from Equations 6.9 and 6.10, Lemma 6.3.13 and Theorem

6.3.6 that

$$\lim_{n \to \infty} \frac{\log \mu_{\infty,\beta}[B_{\delta_n}(x)]}{\log(b^{-n})} \overset{d}{=} \beta(y)y - \theta^\star(\beta(y))$$

$$= \inf_q \{qy - \theta^\star(q)\}$$

$$= f^\star(y).$$

3. Then appeal to Billingsley's Theorem 4.5.6.

(a) Let

$$F_\beta(y) = \left\{ x \in [0,1] : \lim_{n \to \infty} \frac{\log \mu_{\infty,\beta}[B_{\delta_n}(x)]}{\log(b^{-n})} = f^\star(y) \right\}.$$

Since $\mu_{\infty,\beta} F_\beta(y) = \mu_{\infty,\beta}([0,1])$ with probability 1, then

$$\dim_{\mu_{\infty,\beta}} F_\beta(y) = 1.$$

(b) By construction, it then follows that

$$\dim_H \widetilde{F}(y) = \dim_H F_\beta(y)$$

$$= f^\star(y) \dim_{\mu_{\infty,\beta}} F_\beta(y)$$

$$= f^\star(y).$$

$\square$

6.3.15 Proposition (Holley & Waymire, 1992)

If $\widetilde{F}(y) \neq \emptyset$, then $\dim_B \widetilde{F}(y) = 1$. $\square$

The multifractal spectrum, $f^\star(y)$, does not necessarily represent a geometrical dimension over any, let alone its entire range. If the conditions of Theorem 6.3.14 are satisfied, then $f^\star(y)$ can be interpreted as a Hausdorff dimension when $f^\star(y) > 0$. However, there is often a considerable part of the range of $f^\star(y)$ where it is negative (see Example 6.3.16 below). These values have sometimes been referred to in the literature as 'negative dimensions'. There have been a number of papers by Mandelbrot (1989, 1990a, 1991) describing such a phenomenon. Mandelbrot refers to those values of y such that $f^\star(y) > 0$ as *manifest*, and those values of y such that $f^\star(y) < 0$ as *latent*. Further, those values of y where $y < 0$ and $f^\star(y) > -\infty$ are referred to as *virtual*. These cases are most easily seen in the following example.

6.3.16 Example - Log-Normal Distribution

Let the random weight be defined as $W = \exp(-Z)$, where Z is a normal random variable with variance σ^2 and mean $\frac{1}{2}\sigma^2 + \log b$ (see §1.5.2; Kolmogorov, 1962; Oboukhov, 1962; Mandelbrot, 1989 and Meneveau & Sreenivasan, 1991). Then W is a positive random variable with a log-normal distribution, and mean b^{-1}.

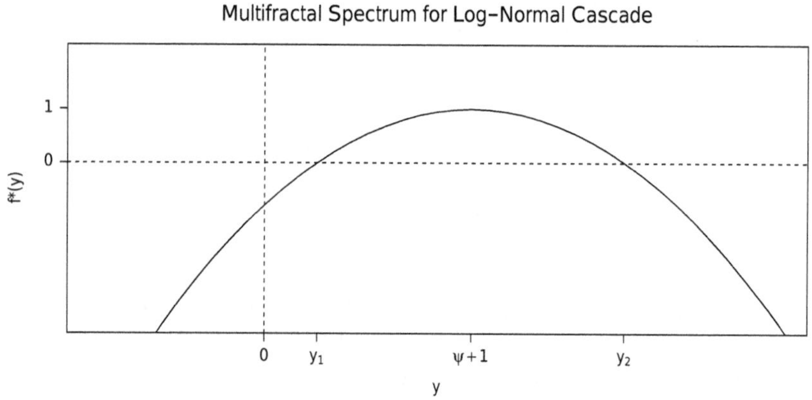

Multifractal Spectrum for Log–Normal Cascade

Figure 6.2 *Multifractal spectrum for the log-normal random cascade process. $f^\star(y)$ is positive for $y_1 < y < y_2$, where $y_1 = \psi + 1 - 2\sqrt{\psi}$ and $y_2 = \psi + 1 + 2\sqrt{\psi}$. Compare with that of the multinomial measure, Figure 3.2, where $\tilde{f}(y) = -\infty$ for all $y < y_1$ and $y > y_2$.*

This is a special case of the log-Lévy generators of Example 6.3.7, where $\xi = \frac{1}{2}\sigma^2 + \log b$, $\alpha = 2$, and $c = \frac{1}{2}$. It follows from Example 6.3.7 that

$$\log \mathrm{E}[W^q] = \frac{\sigma^2 q^2}{2} - \frac{\sigma^2 q}{2} - q \log b.$$

Therefore,

$$\theta^\star(q) = -\psi q^2 + (\psi + 1)q - 1,$$

where $\psi = \sigma^2/(2 \log b) > 0$. Since $\mathrm{E}[W \log_b W] = (\psi - 1)/b$, it follows from Theorem 6.3.3 that $\mathrm{E}[Z_\infty] = 1$ iff $\psi < 1$; i.e., $\mu([0, 1])$ has an expected value of one iff $\psi < 1$.

The multifractal spectrum, $f^\star(y)$, is calculated as $\inf_q \{qy - \theta^\star(q)\}$. The function $\theta^\star(q)$ is quadratic, hence given y, we minimise $qy - \theta^\star(q)$, and solve for q, say q_{crit}. Taking the derivative, and setting equal to zero gives

$$y = -2\psi q_{\mathrm{crit}} + \psi + 1,$$

and rearranging gives

$$q_{\mathrm{crit}} = \beta(y) = \frac{1 + \psi - y}{2\psi}.$$

Inserting back into $f^\star(y) = \beta(y)y - \theta^\star(\beta(y))$ and simplifying, one gets

$$f^\star(y) = 1 - \frac{(y - \psi - 1)^2}{4\psi}.$$

The multifractal spectrum, $f^\star(y)$, is plotted in Figure 6.2. Note that $\psi > 0$, hence $f^\star(y) \to -\infty$ as $y \to \pm\infty$. Further, note that $\theta^\star(q)$ is finite for all $q \in \mathbb{R}$ and $-\infty < f^\star(y) \le 1$ for all $y \in \mathbb{R}$. The is an important difference between the multifractal spectrum plotted here in Figure 6.2 and that in Figure 3.2 for the multinomial measure. In that case both the domain and range of the function $\tilde{f}(y)$ were positive. Here both can also be negative.

The reason for the difference between the multifractal spectrum of the multinomial measure (or more generally the deterministic cascade processes) and the random cascade process is due to the difference in the way that mass is allocated to each subset at the nth stage by the mapping W_i in Equation 6.2. In the case of the deterministic cascade processes, the range of this mapping was in fact $(0, 1)$, and hence was a probability. Further, at each stage, the multiplier W_i had a fixed possible maximum and minimum. These values effectively determine the boundaries of the domain of permissible values of $f^\star(y)$ (see Equations 3.15 and 3.16). By construction, the probability was conserved at each step, and was only rearranged within each subset of $\mathcal{K}_n$.

In the case of the random cascades, the permissible range of W_i was $(0, \infty)$, though with $\mathrm{E}[W] = b^{-1}$. Hence an 'average' random measure will have total mass of one. If W is drawn from a long tailed distribution, then every so often one would sample an extremely large value, possibly even to that extent that the mass in a given subset $J_n(\omega)$ is greater than one, hence $Y_n^\ddagger < 0$. This explains why the domain of $f^\star(y)$ can include negative values. The explanation why $f^\star(y) < 0$ is given by Theorem 6.3.9. It is easiest to interpret the function $I_0(y) = 1 - f^\star(y)$, which is a non-negative quadratic with a minimum of zero, from a probabilistic perspective. For small values of y, $I_0(y)$ describes the likelihood of $Y_n^\ddagger$ having such small values, or equivalently, a subinterval being allocated an extremely large weight. Similarly, for large values of y, $I_0(y)$ describes the likelihood of subintervals being allocated very small weights. Corollary 6.3.10 describes the same characteristics of Y_n, i.e., the local behaviour of the limiting measure μ. $\square$

6.4 Other Cascade Processes

Both cascade constructions considered in §6.2 and §6.3 were based on deterministic subdivisions. Falconer (1986), Graf (1987) and Mauldin & Williams (1988) have shown that similar results also hold in the compact case if the similarity ratios at each step are random. Arbeiter (1991) has considered the non-compact case.

More recently, Falconer (1994) considered the case of statistically self-similar measures, where the subdivisions were not only random as in Falconer (1986), but the allocation of weight was also random as in §6.3. This is also discussed extensively in a monograph by Olsen (1994).

Molchan (1995) analyses zeros of Brownian motion. He does this by using two measures. The local time measure is constructed by eliminating all points on the time axis that are further than δ from a zero. Then define a Lebesgue measure $\lambda_\delta(dt)$ on the remaining intervals. The local time measure is the limit of the normalised measure $c\delta^{-1/2}\lambda_\delta(dt)$ as $\delta \to 0$. In the second case he considers the growth of the number of δ-clusters as $\delta \to 0$. In this example there are elements of both random measure allocation and subdivision. He establishes the functions $\theta^\star(q)$ and $f^\star(y)$ and shows that they are related by the Legendre transform $\theta^\star(q) = \inf_y \{qy - f^\star(y)\}$.

6.4.1 Other Definitions of $\theta^\star(q)$

There are a number of other definitions of $\theta^\star(q)$ that are similar in nature to that in Definition 6.1.2. Halsey et al. (1986) proposed a definition of $\theta^\star(q)$ as follows. Let

$$\mathcal{G}^{q,\xi}(F) = \lim_{\delta \to 0} \inf_{\{U_i\}} \left\{ \sum_{i=1}^{\infty} \mu(U_i)^q |U_i|^{-\xi} : |U_i| \le \delta \text{ and } F \subseteq \bigcup_{i=1}^{\infty} U_i \right\},$$

where the infimum is taken over all coverings of F. Then for each value of q, define $\theta^\star(q)$ as

$$\theta^\star(q) = \inf\{\xi : \mathcal{G}^{q,\xi}(F) = 0\} = \sup\{\xi : \mathcal{G}^{q,\xi}(F) = \infty\}.$$

This case is also discussed by Pesin (1993, page 542) who relates it to various other definitions of generalised dimensions. Note that $\dim_H(F) = -\theta^\star(0)$. This definition accounts for irregular size of covers but not for random measure allocation.

A number of authors have proposed various other definitions for $\theta^\star(q)$ that are variations of the above. Olsen (1994, Chapter 2) gives a definition based on point centred spherical covers, where the centre point is contained in the set F. Olsen (1994) also gives a similar definition based on the packing dimension (see Tricot, 1982 and Falconer, 1990), where F is covered by a packing of spheres with centres in F.

PART III

ESTIMATION OF THE RÉNYI DIMENSIONS

CHAPTER 7

Interpoint Distances of Order q and Intrinsic Bias

7.1 Introduction to Part III

There are a number of reasons why one may want to estimate multifractal characteristics of a probability measure μ. Many of the measures of location, spread and so on that are used to characterise classical probability distributions are not useful when the probability measure contains singularities of possibly many different orders. If a probability measure has singularities, possibly of many different orders, then one possibility is to characterise it on the basis of its singularities. The multifractal spectrum is one way to describe the sizes of subsets of $\mathcal{X}$ containing singularities of a given order. However, the multifractal spectrum is difficult to deal with numerically, and hence one usually estimates the Rényi dimensions, and then calculates the Legendre transform to produce the multifractal spectrum (Chapter 2).

The dimension estimates also describe the degree of non-uniformity of the measure μ on the attracting set of a dynamical system. Often quite simple sets of non-linear equations (non-stochastic) can generate extremely complex behaviour that appears to be stochastic. Dimension estimation has been used to determine whether a time series has been generated by a deterministic process of typically low dimensionality or a stochastic system. It should be noted though, that many stochastic processes also have properties that scale in a powerlaw manner, and some of the concepts and estimation techniques discussed are transferable to these situations too.

The problem with many physical processes, in particular, is that the whole process is not observable. For example, with earthquakes, one may only record characteristics of the seismic waves generated by the event. From this information, some other characteristics of the event are determined. In weather systems measurements like rainfall, wind velocity and cloud cover can be recorded at certain locations. Together with satellite images, it is attempted to put together an overall picture, possibly of regional or even global climate. In some situations, only a scalar time series is recorded. In order to calculate the dimensions of the underlying equations driving the process, a higher dimensional space is reconstructed which can generally be shown to have the same fractal properties as the unobserved system. For much of our discussions in Part III, we will assume that $\mathcal{X} \subseteq \mathbb{R}^d$ is directly observable, and that μ is concentrated on a subset of $\mathcal{X}$.

This is because the notion of embedding and reconstruction adds another level of complication to the discussion which is generally not necessary. Embedding and reconstruction will be discussed in §10.4.4.

In early studies involving fractal dimensions of dynamical systems, dimensions were estimated by covering the d-dimensional space $\mathcal{X}$ with d-dimensional boxes of width δ. The number of visited boxes would then be counted. This would be repeated for a sequence of smaller and smaller values of δ. The slope of the plot of the logarithm of the count by $\log \delta$ would be calculated in a region where the line was straight. This methodology relates to the lattice based multifractals discussed in Chapter 4, in particular $\widetilde{D}_0$.

The number of calculations required to estimate $\widetilde{D}_0$ was large because as δ became sufficiently small, most of the boxes were not even visited. Grassberger & Procaccia (1983a, b, c) suggested an alternative method that eliminated the need to count the visits to boxes, most of which would be zero. They argued that one would get a sufficiently good idea of the dimension by estimating the powerlaw exponent of the probability distribution of pairs of interpoint distances. However, this dimension is D_2 and is part of the family discussed in Chapter 5. Our discussions in Part III are also based on interpoint distances and are therefore related to the multifractal formalism described in Chapter 5.

7.1.1 Overview of Chapters in Part III

It will be shown that the method of Grassberger & Procaccia (1983a, b, c) based on pairs of interpoint distances can be extended to qth order interpoint distances (as in Theorem 2.4.5), which can be used to calculate higher order point centred Rényi dimensions, D_q, for $q = 2, 3, \cdots$. We denote this qth order interpoint distance as Y. Estimating the Rényi dimensions for $q = 2, 3, \cdots$, essentially involves estimating the powerlaw exponent of the probability distribution function, $F_Y(y)$, of Y.

The probability distribution, $F_Y(y)$, of interpoint distances is often not strictly powerlaw in nature. This can occur when the probability measure is supported on a fractal set. However, there is an 'average' powerlaw if the tangent on a log log plot is drawn over a sufficiently long interval. This can be thought of as a form of bias, though it is not caused by sampling or other methodological deficiencies. It is actually an *intrinsic* part of the process. This form of bias is discussed in Chapter 7.

In Chapter 8 a generalisation of the Grassberger-Procaccia method is given for D_q where $q = 2, 3, \cdots$. However, the main emphasis is on a modified version of the Hill estimator and its statistical properties. This estimator follows more naturally from the perspective of maximum likelihood estimation, which may be the first line of attack for a statistician. However, it does have a particular problem with bias that is not so evident when using the Grassberger-Procaccia method. This bias appears to be largely driven by the *intrinsic* bias discussed in Chapter 7.

Most real data only partially describe what is of real interest to the researcher. For example, data are often contaminated with noise. Further, this noise may not be uniform in either time or space. For example, catalogues of earthquake locations will be more accurate in regions with a higher density of seismic stations. Accuracy will also vary over time depending on when stations are actually active. All data are rounded which effectively creates discrete random variables from some underlying continuum. Another form of bias is caused by not observing the entire process, by placing arbitrary boundaries in both time and space. This creates a boundary effect. We refer to these as *extrinsic* forms of bias and discuss their effects in Chapter 9.

Chapter 10 describes some uses of dimension estimation, though applying the techniques to data simulated from mathematical and statistical models rather than real data. Using data with at least partially understood properties enables us to more easily evaluate the methods and interpret the results.

Chapter 11 consists of some case studies using real data. We attempt to disentangle those aspects of the dimension plots that are attributable to various forms of bias and data deficiencies, and those that may have a genuine fractal interpretation.

7.1.2 Review of Notation

In §2.4.1, $\theta(q)$ was defined as

$$\theta(q) = \lim_{\delta \to 0} \frac{\log \left[\int_{\mathcal{X}_\delta} \mu^{q-1} [S_\delta(x)] \mu(dx) \right]}{\log \delta} \qquad -\infty < q < \infty,$$

where $\mathcal{X}_\delta = \{x \in \mathcal{X} : \mu[S_\delta(x)] > 0\}$. We will refer to $\theta(q)$ as the *qth order correlation exponent*, and $\int \mu^{q-1}[S_\delta(x)]\mu(dx)$ as the *qth order correlation integral*. The point centred Rényi dimensions were then defined as

$$D_q = \frac{\theta(q)}{q-1} \qquad (7.1)$$

for $q \neq 1$. $D_2 = \theta(2)$ is often referred to as the *correlation dimension*.

Let $X_1, X_2, \cdots, X_q$ be a sample of independent random variables drawn from the probability distribution μ, and define Y (as in Theorem 2.4.5) as

$$Y = \max\{\|X_1 - X_q\|, \|X_2 - X_q\|, \cdots, \|X_{q-1} - X_q\|\}.$$

We will refer to Y as the *interpoint distance of order q*. Note that Y is always determined by q, though to avoid notation being clumsy, we have not indicated this explicitly in the form of a subscript. In our calculations, unless otherwise stated, $\|\ \|$ always refers to the $\mathcal{L}^\infty$ or max norm. From Theorem 2.4.5, the probability distribution of Y, for $q = 2, 3, 4, \cdots$, is

$$F_Y(y) = \Pr\{Y \leq y\} = \int \mu[S_y(x)]^{q-1}\mu(dx). \qquad (7.2)$$

A corollary to Theorem 2.4.5 is that, for $q = 2, 3, 4, \cdots$, the correlation exponents are

$$\theta(q) = \lim_{y \to 0} \frac{\log F_Y(y)}{\log y}, \tag{7.3}$$

if the limit exists.   □

In this chapter, we investigate the behaviour of the qth order correlation integral, particularly for probability measures that are supported on a self-similar set. The Rényi dimensions are a limiting concept, where interpoint distances become small. Here we look at those *intrinsic* properties of the qth order correlation integral that may affect the properties of estimators of the Rényi dimensions.

7.2 Boundary Effect

Consider the situation where one randomly samples interpoint distances Y. We know from §2.4.3 that if the probability measure μ can be represented by a probability density function, then D_q will be constant for all $q > 0$. Further, the value of D_q will be d, the dimension of the range of the probability density function. From Equations 7.1 and 7.3 it follows, for $q = 2, 3, \cdots$, that

$$D_q = \frac{1}{q-1} \lim_{y \to 0} \frac{\log F_Y(y)}{\log y}.$$

Note that the Rényi dimensions describe the powerlaw exponent of $F_Y(y)$ for infinitely small values of y, and this powerlaw behaviour may not necessarily extend to larger values of y. The boundary effect refers to an apparent deficit of larger interpoint distances relative to that predicted by a strict powerlaw behaviour. As the size of the interpoint distance becomes comparable to that of the width of the region, the chance of sampling such a value of Y decreases.

In this section, we analyse the boundary effect for a few simple situations. In §7.3 it will be shown that fractal like behaviour is related to a multiplicity of these boundary effects.

7.2.1 Distribution of $\mathcal{L}^\infty$ Norm

Let $q \geq 1$ and X_i ($i = 1, 2, \cdots, q$) be independent random vectors in $\mathbb{R}^d$, i.e.,

$$X_i = (X_{i1}, X_{i2}, \cdots, X_{id})'.$$

Further, assume that $X_{i1}, X_{i2}, \cdots, X_{id}$ are independent. Using the $\mathcal{L}^\infty$ norm, Y can be expressed as

$$\begin{aligned} Y &= \max_{1 \leq j \leq q-1} \|X_j - X_q\|_\infty \\ &= \max_{1 \leq j \leq q-1} \max_{1 \leq k \leq d} |X_{jk} - X_{qk}|. \end{aligned}$$

Assume that $|X_{jk} - X_{qk}|$, $\forall j$ and k, has the same distribution as the random variable U. Then

$$F_Y(y) = \Pr\{Y \le y\} = [F_U(y)]^{d(q-1)}, \tag{7.4}$$

and the qth order correlation exponent is

$$\theta(q) = d(q-1) \lim_{y \to 0} \frac{\log F_U(y)}{\log y}.$$

If the density of U exists, then $\theta(q) = d(q-1)$ and $D_q = d$ for $q \ge 2$. □

7.2.2 Example - Gaussian Distribution

Consider the case where $q = 2$ and X_1 and X_2 are multivariate normal random vectors in $\mathbb{R}^d$. Using the $\mathcal{L}^2$ norm, the difference is also normally distributed with twice the variance. Thus

$$
\begin{aligned}
F_Y(y) &= \int \mathbf{1}(\|x\|_2 \le y) \frac{1}{\sqrt{2}} f_X\left(x/\sqrt{2}\right) dx \\
&= \int \mathbf{1}\left(\|\sqrt{2}x\|_2 \le y\right) f_X(x) dx \\
&= \frac{|\Sigma^{-1}|^{1/2}}{(2\pi)^{d/2}} \int \mathbf{1}\left(\|\sqrt{2}x\|_2 \le y\right) \exp\left(-\frac{1}{2}x'\Sigma^{-1}x\right) dx \\
&= \frac{|D^{-1}|^{1/2}}{(2\pi)^{d/2}} \int \mathbf{1}\left(\|\sqrt{2}z\|_2 \le y\right) \exp\left(-\frac{1}{2}z'D^{-1}z\right) dz \\
&\qquad\qquad \text{where } D = P'\Sigma P = \operatorname{diag}(\lambda_1, \cdots, \lambda_d) \\
&= \frac{1}{(2\pi)^{d/2}} \int \mathbf{1}(2w'Dw \le y^2) \exp\left(-\frac{1}{2}w'w\right) dw \\
&\qquad\qquad \text{where } w_i = z_i \lambda_i^{-1/2} \quad i = 1, \cdots, d \\
&= \Pr\left\{ 2\sum_{i=1}^{d} \lambda_i W_i \le y^2 \,\middle|\, W_i \text{ are i.i.d. } \chi_1^2 \right\}.
\end{aligned}
$$

The program by Davies (1980) can be used to calculate quadratic form probabilities.

When $\Sigma = I$, then $F_Y(y)$ can be expressed in terms of the chi-squared distribution with d degrees of freedom, as

$$F_Y(y) = \frac{2^{-d/2}}{\Gamma(d/2)} \int_0^{y^2/2} w^{d/2-1} \exp(-w/2) dw.$$

Using integration by parts, one gets

$$F_Y(y) = y^d \frac{\exp(-y^2/4)}{2^d} \sum_{k=0}^{\infty} \frac{1}{\Gamma(d/2 + k + 1)} \left(\frac{y^2}{4}\right)^k. \tag{7.5}$$

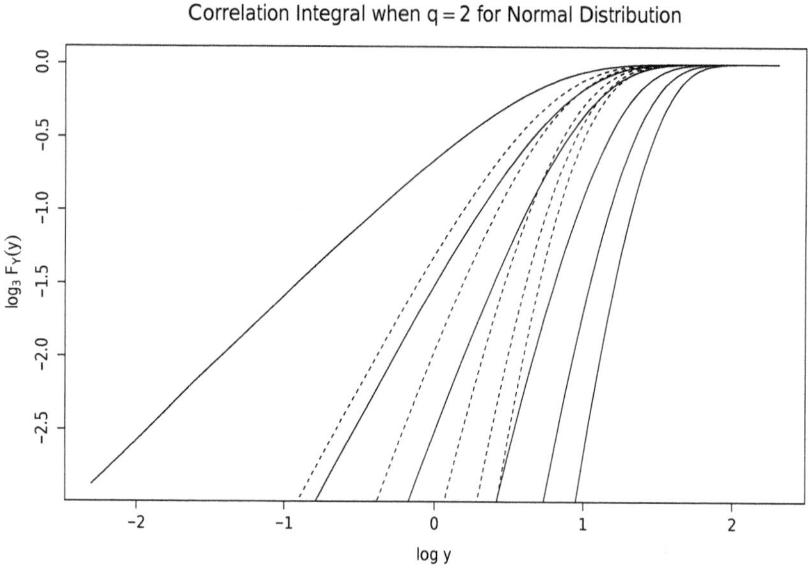

Correlation Integral when q = 2 for Normal Distribution

Figure 7.1 *Correlation integrals when $q = 2$ for the normal distribution in $d = 1$ (top line), 2, 3, 5, 7 and 9 (bottom line) dimensions. Using the $\mathcal{L}^2$ norm produces the solid lines, and the $\mathcal{L}^\infty$ norm produces the dashed lines. Note that in the case of $d = 1$, they are the same. As $y \to 0$ the slope in the $\log y$ vs. $\log F_Y(y)$ plot tends to d.*

It can be seen that $F_Y(y)$ consists of two parts, a powerlaw part with exponent d, and a non-powerlaw part. The non-powerlaw part is monotonically decreasing with limit $[2^d\Gamma(d/2+1)]^{-1}$ as $y \to 0$. It effectively describes the boundary effect which causes a deficit of larger interpoint distances relative to a pure powerlaw occurrence.

Note that by inserting $d = 1$ into Equation 7.5 and using Equation 7.4, an expression for the correlation integral using the $\mathcal{L}^\infty$ norm can also be calculated. A comparison between the $\mathcal{L}^\infty$ and $\mathcal{L}^2$ norms is plotted in Figure 7.1.   □

Takens (1993, page 245) shows that, if different norms only differ by a factor that is bounded, and are bounded away from zero, then the correlation dimension will be the same.

7.2.3 Example - Uniform Distribution

Let $X_1, \cdots, X_q$ be independent continuous uniform random vectors in $[0,1]^d$ with independent components. Then the qth order interpoint distance has probability distribution $F_Y(y) = y^{d(q-1)}(2-y)^{d(q-1)}$ and the correlation exponents

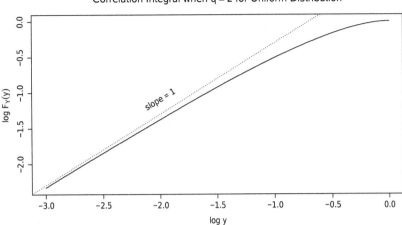

Figure 7.2 *The correlation integral when $q = 2$ for the uniform distribution when $d = 1$. As $y \to 0$, the slope of the line tends to one. The curve in the line as $y \to 1$ is caused by the boundary effect.*

for $q = 2, 3, \cdots$ are

$$\theta(q) = \lim_{y \to 0} \frac{\log \left[y^{d(q-1)} (2 - y)^{d(q-1)} \right]}{\log y} = d(q - 1).$$

Again, as in Example 7.2.2, the probability function of Y has a powerlaw and non-powerlaw part, with the non-powerlaw part describing the boundary effect. □

7.2.4 Example - 'Wrap-Around' Metric

Theiler (1990) showed that by using a 'wrap-around' metric for differences of uniform random variables, the boundary effect is eliminated. That is, let X_1 and X_2 be i.i.d. uniform on $[0, 1]$ (i.e., $d = 1$). Then for $q = 2$,

$$Y = \|X_1 - X_2\| = \begin{cases} |X_1 - X_2| & \text{if } |X_1 - X_2| \leq \frac{1}{2} \\ 1 - |X_1 - X_2| & \text{if } |X_1 - X_2| > \frac{1}{2}. \end{cases}$$

In this case, $F_Y(y) = 2y$ for $0 < y < \frac{1}{2}$, and thus the 'non-powerlaw part' is constant for all $y < \frac{1}{2}$. This is the same idea as uniformly distributed points on the circumference of a circle ($d = 1$), or on the surface of a sphere ($d = 2$) where the distance is taken as the shortest arc length between two points. □

7.3 Multiplicity of Boundaries

In the previous section, the boundary effect was portrayed as a bias, and consequently a nuisance from an estimation perspective. This is not completely the case. In fact, fractal dimensions usually occur because of a multiplicity of boundaries that occur in a powerlaw manner.

In Example 7.2.3, we considered the uniform distribution on the unit interval. The boundary effect was displayed in Figure 7.2. Consider again the uniform distribution, but not necessarily on the unit interval. Let $\mathcal{K}_0$ be the unit interval, $\mathcal{K}_1$ be the unit interval with the middle third cut out, etc., as for the construction of the Cantor set. This situation is the same as was described in Example 1.2.1. Using a value of $p_0 = \frac{1}{2}$ ensures that the allocation of probability to each subinterval is the same, and hence uniform. Further note that $G_n(y)$, given by Equation 3.18, is the probability distribution of the interpoint distances when $q = 2$ on $\mathcal{K}_n$.

The correlation integrals ($q = 2$) for the uniform distribution on $\mathcal{K}_n$, for $n = 0, \cdots, 7$, are plotted in Figure 7.3. The lower solid line is when the uniform distribution is supported on $\mathcal{K}_0$, and hence is the same as that in Figure 7.2. The second solid line from the bottom is when the measure is supported on $\mathcal{K}_1$, etc. The two dotted lines are only reference lines, the upper one having slope $\log_3 2$ and the lower one with slope 1. Note that the correlation dimension, D_2, of the Cantor measure with $p_0 = \frac{1}{2}$ is $\log_3 2$.

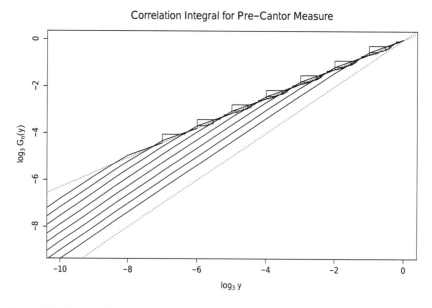

Figure 7.3 *The correlation integral ($q = 2$) for the pre-Cantor measure ($p_0 = 0.5$) for $k = 0, \cdots, 7$. The slope of the lower dotted line is one, and is the asymptote of the slope when the interpoint distances are small. The slope of the upper dotted line is equal to D_2.*

Consider the case for $\mathcal{K}_1$. When interpoint distances are less than $\frac{1}{3}$, we essentially have two separate uniform distributions, both identical, where each pair of points originate from the same subinterval. Hence for small values of y, the correlation integral behaves in the same way as for the uniform distribution with the line having an asymptotic slope of one as $y \to 0$.

However, it is not possible to have interpoint distances y such that $\frac{1}{3} < y < \frac{2}{3}$. Consequently, the probability function $G_1(y)$ will be flat in this region. Once $y \geq \frac{2}{3}$, interpoint distances are again possible, however, each point must come from each different subinterval. When $n = 2$, we add another 'hole' to the support of the measure. This adds another kink to the plot of the correlation integral. Each time a middle third is cut, another kink is added. When the interpoint distances are greater than the smallest 'hole', the slope of the line is roughly D_2. When they are smaller than the smallest 'hole', it behaves like the uniform distribution with slope tending to one.

As n increases, $G_n(y)$ will converge to the distribution function of the Cantor measure. The correlation integral, $F_Y(y)$, will be well approximated by $G_{11}(y)$ for the ranges plotted in Figure 7.4.

We have referred to the cut subintervals as 'holes' in the support of the measure. The oscillatory behaviour on the logarithmic scale shown in Figure 7.4 is often

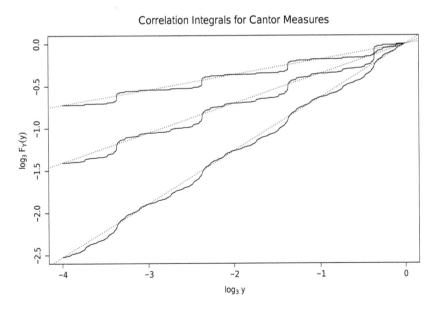

Figure 7.4 *Correlation integrals* ($q = 2$) *for various cases of the Cantor measure where* $p_0 = 1 - p_2$ *and* $p_1 = 0$. *The top line is when* $p_0 = 0.1$, *middle when* $p_0 = 0.2$, *and bottom when* $p_0 = 0.5$. *The dotted lines have slope* $D_2 = -\log_3(p_0^2 + p_2^2)$.

described as lacunary in the literature, the word being derived from the Latin word *lacuna*, meaning hole.

7.4 Decomposition of $F_Y(y)$

In this section we investigate the decomposition of $F_Y(y)$ into powerlaw and non-powerlaw components. We use the notation of Theiler (1988), denoting the non-powerlaw component by $\Phi(y)$.

7.4.1 Proposition

The correlation dimension exists iff the correlation integral can be decomposed into the form

$$F_Y(y) = \Phi(y)y^{\zeta_q}, \tag{7.6}$$

where ζ_q is a positive constant and $\Phi(y)$ is a positive function such that

$$\lim_{y \to 0} \frac{\log \Phi(y)}{\log y} = 0; \tag{7.7}$$

i.e., $|\log \Phi(y)| = o(|\log y|)$ as $y \to 0$. Further, given this decomposition, $\theta(q) = \zeta_q$.

Proof. Equations 7.6 and 7.7 imply that $\theta(q) = \zeta_q$. Conversely, if $\theta(q)$ exists, then

$$\lim_{y \to 0} \frac{\log F_Y(y)}{\log y} = \theta(q).$$

Since

$$F_Y(y) = y^{\theta(q)} \frac{F_Y(y)}{y^{\theta(q)}},$$

then

$$\lim_{y \to 0} \frac{\log(F_Y(y)/y^{\theta(q)})}{\log y} = 0,$$

and therefore $\Phi(y) = y^{-\theta(q)} F_Y(y)$. ◻

7.4.2 Definitions - Self-Similar Type Behaviour

1. $F_Y(y)$ will be said to be *self-similar* with scale parameter ζ_q if for all $s \in [0, 1]$, it satisfies the scaling relation

$$F_Y(sy) = s^{\zeta_q} F_Y(y),$$

for all y such that $y < y_{\max} = \inf\{y : F_Y(y) = 1\}$.

2. $F_Y(y)$ will be said to have *strictly powerlaw behaviour* with scale parameter ζ_q if

$$F_Y(y) = ay^{\zeta_q},$$

for all y such that $y < y_{\max} = \inf\{y : F_Y(y) = 1\}$ and $a = y_{\max}^{-\zeta_q}$.

3. The distribution $F_Y(y)$ will be said to have a *regularly varying lower tail* with exponent ζ_q if for all $s > 0$

$$\lim_{y \to 0} \frac{F_Y(sy)}{F_Y(y)} = s^{\zeta_q}.$$

□

If $F_Y(y)$ has a regularly varying lower tail with exponent ζ_q, then

$$F_Y(sy) = s^{\zeta_q} F_Y(y) + o(F_Y(y))$$

where $o(F_Y(y))/F_Y(y) \to 0$ as $y \to 0$. As such, regular variation is weaker form of self-similarity.

7.4.3 Proposition

If $F_Y(y)$ is self-similar with parameter ζ_q, then $\theta(q)$ exists, and further, $\theta(q) = \zeta_q$.

Proof. By assumption $F_Y(y) = s^{-\zeta_q} F_Y(sy)$, thus by definition

$$\theta(q) = \lim_{y \to 0} \frac{\log F_Y(y)}{\log y} = \lim_{y \to 0} \frac{\log F_Y(sy)}{\log y}.$$

When y is sufficiently small, $F_Y(sy) = y^{\zeta_q} F_Y(s)$. Hence, together with the above equation,

$$\theta(q) = \lim_{y \to 0} \frac{\log[y^{\zeta_q} F_Y(s)]}{\log y} = \zeta_q.$$

□

7.4.4 Proposition

$F_Y(y)$ is self-similar with scale parameter ζ_q iff it has strictly powerlaw behaviour with parameter ζ_q.

Proof. Strictly powerlaw implying self-similarity follows directly from the definitions. Conversely, if $F_Y(y)$ is self-similar, then $\theta(q)$ exists and $\theta(q) = \zeta_q$. The existence of $\theta(q)$ implies that there exists a decomposition of the form $F_Y(y) = \Phi(y)y^{\zeta_q}$. Given self-similarity, $\Phi(sy)(sy)^{\zeta_q} = s^{\zeta_q}\Phi(y)y^{\zeta_q}$, i.e., $\Phi(sy) = \Phi(y)$ for all s between zero and one, and any y. Thus $\Phi(y) = $ constant. □

Feller (1971, page 276) uses a notion of a function being slowly varying at infinity. In the present context, we will say that the function $\Phi(y)$ is slowly varying at zero if, for all $s > 0$,

$$\lim_{y \to 0} \frac{\Phi(sy)}{\Phi(y)} = 1.$$

7.4.5 Proposition

Given that $\theta(q)$ exists, $F_Y(y)$ has a regularly varying lower tail iff $\Phi(y)$ is a slowly varying function at zero.

Proof. Since $\theta(q)$ exists, then there is a decomposition of $F_Y(y)$ into powerlaw and non-powerlaw parts. Also, given that $\Phi(y)$ is a regularly varying function at zero

$$\lim_{y \to 0} \frac{\Phi(sy)(sy)^{\zeta_q}}{\Phi(y)y^{\zeta_q}} = s^{\zeta_q},$$

i.e., $\lim_{y \to 0} \Phi(sy)/\Phi(y) = 1$. The converse is obvious.   □

In the preceding discussion, we have shown that a function $F_Y(y)$ being self-similar is equivalent to it being strictly powerlaw. Such behaviour is too restrictive. We then considered the case where it had self-similar like behaviour for sufficiently small y. Again, this imposes restrictions on the $\Phi(y)$ function that do not hold when the interpoint distances are sampled from a probability distribution whose measure is supported on a fractal set. We require a concept weaker than strict powerlaw behaviour, but where $\theta(q)$ still exists.

7.4.6 Definitions - Lacunary Type Behaviour

1. $F_Y(y)$ will be said to be *lacunary* with scale parameter ζ_q if there exists a constant c between zero and one such that $F_Y(cy) = c^{\zeta_q} F_Y(y)$ for all y such that $y < y_{\max} = \inf\{y : F_Y(y) = 1\}$.   □

2. The distribution $F_Y(y)$ will be said be *lacunary in the lower tail* with exponent ζ_q if there exists a constant c between zero and one such that

$$\lim_{y \to 0} \frac{F_Y(cy)}{F_Y(y)} = c^{\zeta_q}.$$

                    □

Note that lacunarity is a weaker form of self-similarity.

7.4.7 Proposition

If $F_Y(y)$ is lacunary with scale parameter ζ_q, then $\theta(q)$ exists and $\theta(q) = \zeta_q$. Further, $\Phi(y)$ is a bounded periodic function with period one on a logarithmic scale with base c^{-1}.

Proof. $\theta(q) = \zeta_q$ follows in the same manner as for self-similarity. Given lacunarity, there exists a constant c between zero and one, such that

$$\Phi(cy)(cy)^{\zeta_q} = c^{\zeta_q} y^{\zeta_q} \Phi(y),$$

i.e., $\Phi(cy) = \Phi(y)$, so $\Phi(y)$ is periodic with period one on a logarithmic scale with base c^{-1}. Given Equation 7.7, an interval can be selected, say

$$I = (\log_{c^{-1}} y_0, 1 + \log_{c^{-1}} y_0),$$

where y_0 is sufficiently small such that for all $y \in I$, $|\log \Phi(y)| < |\log y|$; i.e., $\log \Phi(y)$ is bounded in I. Since $\Phi(y)$ is periodic, then it must be a bounded function. □

7.4.8 Example

Figure 7.5 shows the non-powerlaw component, $\Phi(y)$, of the probability function $F_Y(y)$, $q = 2$, for the Cantor measure with $p_0 = 0.5$. In fact, the plot is an approximation of $\Phi(y)$ derived after 11 iterations of the recurrence relation given by Equation 3.18. Note that the oscillatory behaviour of $\Phi(y)$ has a constant period on a logarithmic scale, and a constant amplitude. It is this part of the correlation integral that describes clustering properties and is lacunary in the sense of Definition 7.4.6. □

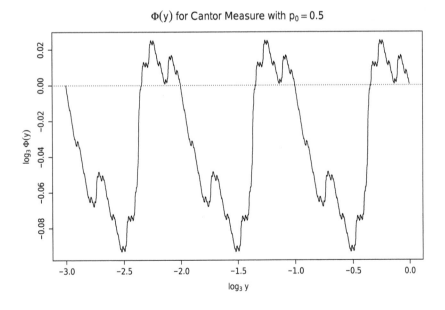

$\Phi(y)$ for Cantor Measure with $p_0 = 0.5$

Figure 7.5 The function $\Phi(y)$ when $q = 2$ for the Cantor measure with $p_0 = 1 - p_2 = 0.5$ and $p_1 = 0$. This is the non-powerlaw component of $F_Y(y)$ when $q = 2$.

In the Cantor measure examples discussed in this chapter, the function $\Phi(y)$ is periodic on a logarithmic scale of base 3 over its entire range, and hence $F_Y(y)$ is lacunary as in Definition 7.4.6. This appears to hold for all $q = 2, 3, \cdots$. Unfortunately, this lacunarity does not automatically hold for all measures supported on self-similar sets. On the basis of our simulations, it appears to be dependent on the placement of 'gaps'. In §10.2, examples will be given of particular cases of the multinomial measure where the function $\Phi(y)$ not only contains a lacunary like component, but also a boundary effect component. Hence $F_Y(y)$ in these cases is only lacunary in the lower tail (see Definition 7.4.6).

7.5 Differentiable Distributions

The Grassberger-Procaccia method of estimating the correlation dimension, to be discussed in the next chapter, involves plotting an estimate of $\log_b F_Y(y)$ against $\log_b y$ for some number b, and estimating the slope. Let $w = \log_b y$. Then this plot is similar to taking the derivative of $\log_b F_Y(b^w)$ with respect to w, i.e., assuming $F_Y(y)$ is differentiable,

$$\frac{d}{dw} \log_b F_Y(b^w) = \theta(q) + b^w \frac{\Phi'(b^w)}{\Phi(b^w)}.$$

If $\Phi(y)$ is monotonically decreasing, the above derivative will be less than $\theta(q)$. This occurs in both Examples 7.2.2 and 7.2.3 and is indicative of a deficit of larger interpoint distances with respect to a powerlaw behaviour. Smith (1992a, b) contains further analysis of the case where $F_Y(y)$, for $q = 2$, is differentiable.

CHAPTER 8

Estimation of Point Centred Rényi Dimensions with $q \geq 2$

8.1 Introduction

In this chapter, we investigate methods of estimating the point centred Rényi dimensions, D_q, for $q = 2, 3, \cdots$, and some of their statistical properties. The methods are based on the qth order interpoint difference, which has been discussed in Chapter 7.

In many time series and dynamical systems contexts, only a scalar process is observed. From this sequence of scalar observations a reconstructed phase space is usually formed. This will not be discussed until §10.4.4. In this chapter we always assume that the series $\{X_i\}$ is observed in $\mathbb{R}^d$, and that the measure μ is supported on a set of dimension less than or equal to d.

8.1.1 Review of Notation

Let $X_1, X_2, \cdots, X_q$ be a sample of independent random variables drawn from the probability distribution μ, and define Y as

$$Y = \max\{\|X_1 - X_q\|, \|X_2 - X_q\|, \cdots, \|X_{q-1} - X_q\|\}.$$

In our calculations, unless otherwise stated, $\| \ \|$ always refers to the $\mathcal{L}^{\infty}$ or max norm. As in Chapter 7 reference to q will not be explicitly stated. The probability distribution function of Y, $F_Y(y)$, is given by Equation 7.2. For $q = 2, 3, 4, \cdots$ the correlation exponents, $\theta(q)$, are given by Equation 7.3 and the Rényi dimensions, $D_q = \theta(q)/(q-1)$, by Equation 7.1.

As for the random variable Y, $\theta(q)$ will generally be written simply as θ to avoid notation becoming clumsy. An estimator of θ will be denoted by Θ, and a specific estimate as $\hat{\theta}$. The estimator Θ will be said to be *unbiased* if $E[\Theta] = \theta$, and *weakly consistent* if for any $\epsilon > 0$, $\lim_{N \to \infty} \Pr\{|\Theta - \theta| > \epsilon\} = 0$, where N is the sample size. Sufficient conditions for weak consistency are that Θ is unbiased and $E[(\Theta - \theta)^2] \to 0$ as $N \to \infty$.

8.1.2 Methods of Estimation

The most commonly used method of estimating D_2 is one devised by Grassberger & Procaccia (1983a, b, c). It is based on the empirical probability distribution of pairs of interpoint differences. The method appeals to the property that the plot of $\log y$ by $\log F_Y(y)$ should roughly be a straight line in some region where the function $F_Y(y)$ exhibits powerlaw behaviour. A suitable region needs to be determined, and then the slope of the line is used as an estimate of the correlation dimension. This method will be reviewed in §8.2.

The second class of estimators is based on the theory of maximum likelihood assuming that $F_Y(y)$ is powerlaw in some sense. In the present context, we are interested in powerlaw behaviour for small values of y. Briefly consider the trivial case where $F_Y(y) = y^\theta$, for $0 < y < 1$; i.e., for all values of y. Given a sample of n independent interpoint distances, the log-likelihood equation is given by

$$L(\theta; y_1, \cdots, y_n) = n \log \theta + (\theta - 1) \sum_{i=1}^{n} \log y_i,$$

yielding a maximum likelihood estimate for θ of

$$\widehat{\theta} = \frac{n}{-\sum_{i=1}^{n} \log y_i}.$$

Note the relationship with the exponential distribution. The random variable $W = -\log Y$ has density $\theta \exp(-\theta w)$.

More generally, assume that $F_Y(y) = ay^\theta$ for sufficiently small values of y, say $y \leq \epsilon$, and a is some positive constant. That is, we are assuming that the distribution is powerlaw in the lower tail without assuming any further global parametric form for the distribution function. Thus this method of estimating the Rényi dimensions is similar to the methodology used in extreme value theory; see for example, Embrechts et al. (1997). A sample of interpoint distances, $y_1, \cdots, y_n$, are collected. We consider two different conditioning regimes.

1. From the sample $y_1, \cdots, y_n$, we condition on those $N(\epsilon)$ values that are less than ϵ. Here $N(\epsilon)$ is a binomial random variable with parameters n and $a\epsilon^\theta$. Using this strategy, one gets the estimator proposed by Takens (1985), and will be referred to as the Takens estimator. This is discussed further in §8.3.

2. The sample is sorted to form order statistics $y_{(1)} \leq y_{(2)} \leq \cdots \leq y_{(m)} \leq \cdots \leq y_{(n)}$. We then condition on $y_{(m)} \leq \epsilon$. This follows the method of Hill (1975), who wanted to estimate the powerlaw exponent in heavy tailed distributions for extreme tail events, and will be referred to as the Hill estimator. This is discussed further in §8.4.

Many observed processes are correlated in time and hence the assumption of sampling independent interpoint differences can be a problem, particularly in the Grassberger-Procaccia algorithm where all possible differences are often included in the calculations. Physicists often thin the time series, taking every mth observation, where m is sufficiently large. Alternatively, one could randomly sample interpoint distances, but take many bootstrap samples. This method will be discussed in §8.5. In the last section we discuss the relative advantages and disadvantages of the different methods.

8.2 Generalised Grassberger-Procaccia Algorithm

8.2.1 The Case When $q = 2$

Grassberger & Procaccia (1983a) considered the case where $q = 2$. Given a finite sequence of vector random variables $X_1, X_2, \cdots, X_N$ in $\mathbb{R}^d$, their method involves calculating all possible interpoint distances, and using these to form an empirical distribution function as an estimate of $F_Y(y)$. That is, for $q = 2$,

$$\widehat{F}_Y^{\text{GP}}(y, N) = \frac{2}{N(N-1)} \sum_{i=1}^{N-1} \sum_{j=i+1}^{N} 1(\|X_i - X_j\| \leq y), \qquad (8.1)$$

where $1(A)$ is one if A is true and zero otherwise. The procedure then involves plotting $\log y$ by sample values of $\log \widehat{F}_Y^{\text{GP}}(y, N)$, and using the slope of the line in some suitable region as an estimate of the correlation dimension $D_2 = \theta(2)$.

8.2.2 The Case When $q = 2, 3, \cdots$

Given a finite sequence of vector random variables $X_1, X_2, \cdots, X_N$ in $\mathbb{R}^d$, define the random function $\widehat{F}_Y^{\text{GP}}(y, N)$ as

$$\widehat{F}_Y^{\text{GP}}(y, N) = \frac{1}{m} \sum_{k=1}^{m} 1(Y_k \leq y), \qquad (8.2)$$

where $Y_1, \cdots, Y_m$ are all possible permutations of the qth order difference given a sample of N points, the kth being

$$Y_k = \max \left\{ \|X_{k_1} - X_{k_q}\|, \|X_{k_2} - X_{k_q}\|, \cdots, \|X_{k_{q-1}} - X_{k_q}\| \right\}.$$

In the situation where $q > 2$, one samples X_{k_q} from $X_1, X_2, \cdots, X_N$ *without* replacement. Then one samples $X_{k_i}, i = 1, \cdots, q - 1$ *with* replacement. In this second phase, there are N^{q-1} possibilities, and in the first phase there are N possibilities. Therefore $m = N(N-1)^{q-1}$.

The function $\widehat{F}_Y^{\text{GP}}(y, N)$ is an estimator of $F_Y(y)$ in Equation 7.2.

8.2.3 The Exponent ν

The correlation dimension ($q = 2$) is often defined differently (Theiler, 1986, Equation 4) as ν, where

$$\nu = \lim_{y \to 0} \lim_{N \to \infty} \frac{\log \widehat{F}_Y^{\text{GP}}(y, N)}{\log y}.$$

When $q = 2$, the correlation exponent, $\theta(2)$, is the same as the corresponding Rényi dimension D_2. We refer to D_2 as the correlation dimension. The limit ν has a form more similar to an estimator, and will only be equal to D_2 as in Definition 7.3 if a suitable law of large numbers exists. The interpretation of ν when such a law of large numbers does not exist is somewhat questionable.

8.2.4 Determination of the Slope

The determination of the interval where the slope of the line should be estimated is non-trivial. Ideally, one wants to select an interval (y_1, y_2), where $y_{min} \leq y_1 < y_2 \leq y_{max}$, and where a chord on the curve has the same long range slope as $\log F_Y(y)$, then

$$\widehat{\theta}(q) = \text{estimated slope} = \frac{\log \widehat{F}_Y^{\text{GP}}(y_2, N) - \log \widehat{F}_Y^{\text{GP}}(y_1, N)}{\log(y_2) - \log(y_1)}.$$

The strategy that one uses is quite dependent on the behaviour of $\Phi(y)$. The characteristics of $\Phi(y)$ may not be initially known.

In the case where $\Phi(y)$ is monotonically decreasing, there is a deficit of larger interpoint distances, which can be caused by a boundary effect. Nerenberg & Essex (1990), Essex & Nerenberg (1991) and others have pointed out that boundary effects and the finite nature of the datasets affect values of $\widehat{F}_Y^{\text{GP}}(y, N)$. By targeting smaller values of y, the boundary has little effect, though the smaller bin counts cause greater stochastic error. Noise in the data also causes problems for smaller values of y. Both biases caused by boundaries and noise in data are discussed further in Chapter 9.

In the case of lacunarity with no overall boundary effect, there is not the necessity to sample small y, but a need to draw a chord through a sufficient length of the curve so as not to be misled by periodicity. Further discussion is given by Eckmann & Ruelle (1992) and Ruelle (1990).

8.2.5 Consistency for Ergodic Processes

Dynamical systems were initially mentioned in §1.3. Consistency is discussed, amongst others, by Cutler (1991), Denker & Keller (1986), and Pesin (1993). Generally one can represent a process $(\mathcal{X}, T_\xi, \mu)$ by a mapping $T_\xi : \mathcal{X} \to \mathcal{X}$, where μ is a probability measure on $\mathcal{B}(\mathcal{X})$, and the mapping preserves μ, i.e., $\mu = T_\xi^{-1}\mu$; equivalently μ is invariant to T_ξ. Such systems are assumed to be ergodic, so that time averages converge μ-almost all $x(t_0) \in \mathcal{X}$, where $x(t_0)$ is the initial value in the sequence. The process is often required to be mixing, i.e., for all Borel sets $A, B \subseteq \mathcal{X}$,

$$\lim_{n \to \infty} \mu[A \cap S^{-n}(B)] = \mu(A)\mu(B).$$

Pesin (1993) showed that if the sequence is generated as $x(t_n) = T_\xi^n(x(t_0))$, $n = 1, 2, \cdots$, and the measure μ is ergodic, then for μ-almost every $x(t_0) \in \mathcal{X}$,

$$\lim_{N \to \infty} \widehat{F}_Y^{\text{GP}}(y, N) = F_Y(y).$$

Denker & Keller (1986) showed that if $(\mathcal{X}, T_\xi, \mu)$ is a smooth ergodic dynamical system that satisfies certain mixing conditions, then for $y > 0$, $\widehat{F}_Y^{\text{GP}}(y, N)$ and $\log \widehat{F}_Y^{\text{GP}}(y, N)$ converge in probability to $F_Y(y)$ and $\log F_Y(y)$, respectively.

They also showed that $\widehat{F}_Y^{\mathrm{GP}}(y, N)$ has an asymptotic normal distribution as $N \to \infty$. Cutler (1991) discusses problems and properties of least squares estimators formed by regressing $\log \widehat{F}_Y^{\mathrm{GP}}(y, N)$ on $\log y$.

Mikosch & Wang (1993) showed that if $\{X_i\}$ is a stationary ergodic sequence of d-dimensional random vectors with continuous distribution μ, coinciding with the invariant probability measure of the sequence, then $\widehat{F}_Y^{\mathrm{GP}}(y, N) \xrightarrow{\text{a.s.}} F_Y(y)$ as $N \to \infty$.

8.3 Takens Estimator

Initially assume that the y_i's are independent and $F_Y(y) = ay^\theta$, $a > 0$. From the sample $y_1, \cdots, y_n$, we condition on those $N(\epsilon)$ values that are less than ϵ. Thus the conditional distribution function is given by

$$F_{Y|Y<\epsilon}(y) = \frac{F_Y(y)}{F_Y(\epsilon)} = \left(\frac{y}{\epsilon}\right)^\theta.$$

The log-likelihood equation is given by

$$L(\theta; y_1, \cdots, y_{N(\epsilon)}) = N(\epsilon) \log \theta - N(\epsilon) \log \epsilon + (\theta - 1) \sum_{i=1}^{N(\epsilon)} \log\left(\frac{y_i}{\epsilon}\right).$$

The corresponding maximum likelihood estimator of θ is

$$\Theta_\epsilon = \frac{N(\epsilon)}{-\sum_{i=1}^{N(\epsilon)} \log\left(\frac{Y_i}{\epsilon}\right)}.$$

This estimator was suggested by Takens (1985). Note that it is conditional on $Y_i < \epsilon$.

We are interested in determining the behaviour of the estimator Θ_ϵ when $F_Y(y)$ takes on the general form $F_Y(y) = \Phi(y)y^\theta$.

8.3.1 Proposition (Theiler, 1988)

The estimator $1/\Theta_\epsilon$ is asymptotically unbiased for $1/\theta$ if $\lim_{\epsilon \to 0} \gamma(\epsilon) = 0$, where

$$\gamma(\epsilon) = \int_0^\infty \left[\frac{\Phi(\epsilon e^{-w})}{\Phi(\epsilon)} - 1\right] \theta e^{-\theta w} \, dw.$$

This will occur if $\Phi(y) \to$ const as $y \to 0$. Explicitly,

$$\mathrm{E}\left[\frac{1}{\Theta_\epsilon} \middle| Y \le \epsilon\right] = \frac{1}{\theta}(1 + \gamma(\epsilon)). \tag{8.3}$$

Proof. Let

$$W_i = -\log\left(\frac{Y_i}{\epsilon}\right),$$

then $W_i > 0$ if and only if $Y_i < \epsilon$. It follows that

$$\Pr\{W_i > w | W_i > 0\} = \frac{\Pr\{W_i > w\}}{\Pr\{W_i > 0\}}$$

$$= \frac{F_Y(\epsilon e^{-w})}{F_Y(\epsilon)}$$

$$= \frac{\Phi(\epsilon e^{-w})}{\Phi(\epsilon)} e^{-\theta w}.$$

Using integration by parts, $E[W] = \int w f_W(w)\,dw = \int [1 - F_W(w)]\,dw$ if $F_W(0) = 0$. Thus, by direct integration

$$E\left[\frac{1}{\Theta_\epsilon}\middle| Y \leq \epsilon\right] = E[W_i | W_i > 0]$$

$$= \frac{1}{\Phi(\epsilon)} \int_0^\infty \Phi(\epsilon e^{-w}) e^{-\theta w}\,dw$$

$$= \frac{1}{\theta}\left(1 + \int_0^\infty \left[\frac{\Phi(\epsilon e^{-w})}{\Phi(\epsilon)} - 1\right]\theta e^{-\theta w}\,dw\right).$$

$\square$

Theiler (1988, 1990) uses a more explicit relationship with the correlation integral. By substituting $y = \epsilon e^{-w}$ into the above,

$$E\left[\frac{1}{\Theta_\epsilon}\middle| Y \leq \epsilon\right] = \frac{1}{F_Y(\epsilon)} \int_0^\epsilon \frac{F_Y(y)}{y}\,dy.$$

Using integration by parts gives a series expansion for $\gamma(\epsilon)$ as

$$\gamma(\epsilon) = \frac{1}{\Phi(\epsilon)} \sum_{k=1} (-\epsilon)^k \frac{\Gamma(\theta + 1)}{\Gamma(\theta + k + 1)} \left[\Phi^{(k)}(\epsilon) - \lim_{w \to \infty} \Phi^{(k)}(\epsilon e^{-w}) e^{-(\theta+k)w}\right].$$

If the kth derivative of $\Phi(y)$, denoted by $\Phi^{(k)}(y)$, is bounded for small y in such a way that $\lim_{y \to 0} \Phi^{(k)}(\epsilon y) y^{\theta+k} \to 0$, then

$$\gamma(\epsilon) = \frac{1}{\Phi(\epsilon)} \sum_{k=1} (-1)^k \epsilon^k \frac{\Gamma(\theta + 1)}{\Gamma(\theta + k + 1)} \Phi^{(k)}(\epsilon).$$

If the derivatives vanish, the summation will be finite.

Interest should really focus on $E[\Theta_\epsilon | Y \leq \epsilon]$ not $E[1/\Theta_\epsilon | Y \leq \epsilon]$. The difference can be seen in the following example.

8.3.2 Example

Consider a sequence of independent identically distributed exponential random variables $W_1, \cdots, W_m$ such that $E[W_1] = 1/\theta$. The maximum likelihood estimator of θ is

$$\Theta = \frac{m}{\sum_{i=1}^m W_i}.$$

It is obvious that $E[1/\Theta] = 1/\theta$, however,

$$E[\Theta] = \frac{m}{m-1}\theta.$$

$\square$

8.3.3 Proposition

Given that $F_Y(y) = \Phi(y)y^\theta$, and $E[(\Theta_\epsilon - \theta)^k | Y < \epsilon] < \infty$ for all $k > 1$, then

$$E[\Theta_\epsilon | Y < \epsilon] = \theta(1 - \gamma(\epsilon)) + \frac{E[(\Theta_\epsilon - \theta)^2 | Y < \epsilon]}{\theta} - \cdots.$$

Proof. Consider a Taylor series expansion of $1/\Theta_\epsilon$ about $1/\theta$, then take expectations to give

$$E\left[\frac{1}{\Theta_\epsilon} \middle| Y < \epsilon\right] = \frac{1}{\theta} - \frac{E[(\Theta_\epsilon - \theta)|Y < \epsilon]}{\theta^2} + \frac{E[(\Theta_\epsilon - \theta)^2 | Y < \epsilon]}{\theta^3} - \cdots.$$

Rearranging, the result follows.

$\square$

8.3.4 Example - Uniform Random Variables

In Example 7.2.3, it was shown that the correlation integral for points with a uniform distribution in a d-dimensional unit cube is

$$F_Y(y) = y^d(2 - y)^d,$$

therefore, for $k = 1, \cdots, d$,

$$\Phi^{(k)}(y) = (-1)^k \frac{\Gamma(d+1)}{\Gamma(d-k+1)}(2-y)^{d-k}.$$

It then follows from above that

$$\gamma(\epsilon) = \sum_{k=1}^{d} \frac{\Gamma(d+1)}{\Gamma(d+k+1)} \frac{\Gamma(d+1)}{\Gamma(d-k+1)} \left(\frac{\epsilon}{2-\epsilon}\right)^k.$$

$\square$

Pisarenko & Pisarenko (1995) derive an expression for the variance of the Takens estimator when $q = 2$, and describe its properties.

8.4 Hill Estimator

This estimator is based on the work by Hill (1975). Let Y denote the interpoint distance of order q. Consider a sample of n independent interpoint distances, and let $Y_{(m)}$ be the mth order statistic, $1 \leq m \leq n$; i.e.,

$$Y_{(1)} \leq Y_{(2)} \leq \cdots \leq Y_{(m)} \leq \cdots \leq Y_{(n)}.$$

Here we condition on $Y_{(m)} < \epsilon$.

8.4.1 Distribution of Order Statistics

The distribution function of the mth order statistic is

$$
\begin{aligned}
F_{Y_{(m)}}(y) &= \Pr\{Y_{(m)} \leq y\} \\
&= \Pr\{\text{at least } m \text{ of the } Y\text{'s are } \leq y\} \\
&= \sum_{r=m}^{n} \binom{n}{r} F_Y^r(y)[1 - F_Y(y)]^{n-r} \\
&= \frac{\Gamma(n+1)}{\Gamma(m)\Gamma(n-m+1)} \int_0^{F_Y(y)} t^{m-1}(1-t)^{n-m}\, dt \\
&= n \binom{n-1}{m-1} \int_0^y F_Y^{m-1}(r)[1 - F_Y(r)]^{n-m} f_Y(r)\, dr.
\end{aligned}
$$

It follows that the density function of $Y_{(m)}$ is

$$
f_{Y_{(m)}}(y) = n \binom{n-1}{m-1} F_Y^{m-1}(y)[1 - F_Y(y)]^{n-m} f_Y(y).
$$

From David (1970, Equation 2.2.2), the joint density of the first m order statistics is

$$
f_{Y_{(1)}Y_{(2)}\cdots Y_{(m)}}(y_1, \cdots, y_m) = \frac{n!}{(n-m)!} f_Y(y_1) \cdots f_Y(y_m)[1 - F_Y(y_m)]^{n-m},
$$

and thus the conditional density is

$$
\begin{aligned}
f_{Y_{(1)}\cdots Y_{(m-1)}|Y_{(m)}}(y_1, \cdots, y_{m-1}|y_m) &= \frac{f_{Y_{(1)}Y_{(2)}\cdots Y_{(m)}}(y_1, \cdots, y_m)}{f_{Y_{(m)}}(y_m)} \\
&= \frac{(m-1)!}{F_Y^{m-1}(y_m)} f_Y(y_1) \cdots f_Y(y_{m-1}).
\end{aligned}
$$

$\square$

8.4.2 Maximum Likelihood Estimator

Assume that the correlation function has the form $F_Y(y) = ay^\theta$ for $y \leq \epsilon$. Consider a sample of n ordered random variables $Y_{(1)} \leq Y_{(2)} \leq \cdots \leq Y_{(m)} \leq \cdots \leq Y_{(n)}$. Then an estimator for θ, denoted by Θ_m, is

$$
\Theta_m = \frac{m-1}{-\sum_{i=1}^{m-1} \log\left(\frac{Y_{(i)}}{Y_{(m)}}\right)}. \tag{8.4}
$$

Proof. Consider a sample of n order statistics

$$
y_{(1)} \leq y_{(2)} \leq \cdots \leq y_{(m)} \leq \cdots \leq y_{(n)}.
$$

Since $f_Y(y) = a\theta y^{\theta-1}$, the conditional log-likelihood equation is given by

$L(\theta; y_{(1)}, \cdots, y_{(m)})$

$$= \log(m-1)! - (m-1)\log\left(ay_{(m)}^{\theta}\right) + \sum_{i=1}^{m-1} \log\left(a\theta y_{(i)}^{\theta-1}\right)$$

$$= \log(m-1)! - \theta(m-1)\log y_{(m)} + (m-1)\log\theta + (\theta-1)\sum_{i=1}^{m-1} \log y_{(i)}.$$

The derivative is given by

$$\frac{dL(\theta; y_{(1)}, \cdots, y_{(m)})}{d\theta} = \frac{m-1}{\theta} + \sum_{i=1}^{m-1} \log\left(\frac{y_{(i)}}{y_{(m)}}\right).$$

The estimator follows by setting the derivative equal to zero and solving in the usual way. ☐

8.4.3 Proposition

Given that $F_Y(y) = ay^{\theta}$, where a is a positive constant, then

$$E\left[\frac{1}{\Theta_m}\,\middle|\, Y_{(m)}\right] = \frac{1}{\theta},$$

$$E\left[\Theta_m | Y_{(m)}\right] = \frac{(m-1)\theta}{m-2}$$

and

$$\text{Var}\left(\Theta_m | Y_{(m)}\right) = \left(\frac{m-1}{m-2}\right)^2 \frac{\theta^2}{m-3}.$$

Proof.

$E\left[\frac{1}{\Theta_m}\,\middle|\, Y_{(m)} = y_m\right]$

$$= \int_0^{y_m} \cdots \int_0^{y_2} \frac{-1}{m-1} \sum_{i=1}^{m-1} \log\left(\frac{y_i}{y_m}\right)$$
$$\times f_{Y_{(1)}\cdots Y_{(m-1)}|Y_{(m)}}(y_1, \cdots, y_{m-1}|y_m)dy_1 \cdots dy_{m-1}$$

$$= -\int_0^{y_m} \cdots \int_0^{y_2} \frac{(m-1)!}{m-1} \sum_{i=1}^{m-1} \log\left(\frac{y_i}{y_m}\right) \left(\frac{\theta}{y_m}\right)^{m-1}$$
$$\times \left(\frac{y_1}{y_m}\right)^{\theta-1} \cdots \left(\frac{y_{m-1}}{y_m}\right)^{\theta-1} dy_1 \cdots dy_{m-1}.$$

By substituting $w_i = -\log(y_i/y_m)$, we get

$$
E\left[\frac{1}{\Theta_m}\bigg| Y_{(m)} = y_m\right]
$$

$$
= \frac{(m-1)!}{m-1}\int_0^\infty \cdots \int_{w_2}^\infty \left(\sum_{i=1}^{m-1} w_i\right)\theta^{m-1}
$$

$$
\times \exp\left(-\theta \sum_{i=1}^{m-1} w_i\right) dw_1 \cdots dw_{m-1}
$$

$$
= \frac{(m-1)!}{m-1}\int_0^\infty \cdots \int_{w_2}^\infty \left(\sum_{i=1}^{m-1} i(w_i - w_{i+1})\right)\theta^{m-1}
$$

$$
\times \exp\left(-\theta \sum_{i=1}^{m-1} i(w_i - w_{i+1})\right) dw_1 \cdots dw_{m-1}
$$

$$
= \frac{1}{m-1}\int_0^\infty \cdots \int_0^\infty \left(\sum_{i=1}^{m-1} z_i\right)\theta^{m-1}\exp\left(-\theta \sum_{i=1}^{m-1} z_i\right) dz_1 \cdots dz_{m-1}
$$

$$
= \frac{1}{\theta}.
$$

In the second to last step,

$$
z_i = i(w_i - w_{i+1}),
$$

for $i = 1, \cdots, m-1$, was substituted. Note that

$$
\sum_{i=1}^{m-1} z_i = \sum_{i=1}^{m-1} w_i,
$$

because, by definition, $w_m = 0$. Therefore,

$$
\begin{aligned}
w_{m-1} &= \frac{z_{m-1}}{m-1}\\
w_{m-2} &= \frac{z_{m-2}}{m-2} + \frac{z_{m-1}}{m-1}\\
&\ \ \vdots\\
w_1 &= \frac{z_1}{1} + \frac{z_2}{2} + \cdots + \frac{z_{m-1}}{m-1},
\end{aligned}
$$

and thus the Jacobian is given by

$$
J = \left|\left(\frac{\partial w_i}{\partial z_j}\right)_{ij}\right| = \frac{1}{(m-1)!}.
$$

Using the same substitutions as above, one gets

$$E\left[\Theta_m | Y_{(m)}\right]$$

$$= E\left[\frac{m-1}{-\sum_{i=1}^{m-1} \log\left(\frac{Y_{(i)}}{Y_{(m)}}\right)} \middle| Y_{(m)}\right]$$

$$= \int_0^\infty \cdots \int_0^\infty \frac{m-1}{z_1 + \cdots + z_{m-1}} \theta^{m-1} \exp\left(-\theta \sum_{i=1}^{m-1} z_i\right) dz_1 \cdots dz_{m-1}$$

$$= E\left[\frac{m-1}{Z_1 + \cdots + Z_{m-1}} \middle| Z_i \text{ are i.i.d. exponential with parameter } \theta\right]$$

$$= (m-1) \int_0^\infty \frac{1}{w} \frac{\theta(\theta w)^{m-2}}{\Gamma(m-1)} \exp(-\theta w) \, dw$$

$$= \frac{(m-1)\theta}{m-2}.$$

Appealing to the gamma distribution in a similar manner, one also gets

$$\mathrm{Var}\left(\Theta_m | Y_{(m)}\right) = \left(\frac{m-1}{m-2}\right)^2 \frac{\theta^2}{m-3}.$$

□

We now apply similar transformations to deduce the bias for the more general case where $F_Y(y) = \Phi(y) y^\theta$.

8.4.4 Proposition

Given that $F_Y(y) = \Phi(y) y^\theta$, then

$$E\left[\frac{1}{\Theta_m} \middle| Y_{(m)} = y\right] = \frac{1}{\theta}(1 + \gamma_m(y)), \tag{8.5}$$

where

$$\gamma_m(y) = E\left[\frac{1}{m-1} \sum_{i=1}^{m-1} \log\left(\frac{\Phi(Y_{(i)})}{\Phi(Y_{(m)})}\right) \middle| Y_{(m)} = y\right].$$

Further, given that $E\left[(\Theta_m - \theta)^k | Y_{(m)}\right] < \infty$ for all $k \geq 1$, then

$$E\left[\Theta_m | Y_{(m)} = y\right] = \theta(1 - \gamma_m(y)) + \frac{E\left[(\Theta_m - \theta)^2 | Y_{(m)} = y\right]}{\theta} - \cdots . \tag{8.6}$$

Proof. Let $Y_{(i)}$ denote the ith order statistic of differences, where

$$Y_{(1)} \leq Y_{(2)} \leq \cdots \leq Y_{(m)}.$$

Then

$$\frac{F_Y(Y_{(1)})}{F_Y(Y_{(m)})} \leq \frac{F_Y(Y_{(2)})}{F_Y(Y_{(m)})} \leq \cdots \leq \frac{F_Y(Y_{(m-1)})}{F_Y(Y_{(m)})}$$

are a set of order statistics drawn from a uniform distribution on $[0, 1]$. By taking a log transformation as

$$W_{(i)} = -\log\left(\frac{F_Y(Y_{(i)})}{F_Y(Y_{(m)})}\right)$$

for $i = 1, \cdots, m$, the order statistics

$$W_{(1)} \geq W_{(2)} \geq \cdots \geq W_{(m)} = 0$$

are drawn from an exponential distribution with mean one. Now consider a further transformation

$$Z_i = i\left(W_{(i)} - W_{(i+1)}\right).$$

The Z_i's are a set of independent and identically distributed exponential variables with mean one. Further, it can be seen that

$$\sum_{i=1}^{m-1} Z_i = \sum_{i=1}^{m-1} W_{(i)} = -\sum_{i=1}^{m-1} \log\left(\frac{F_Y(Y_{(i)})}{F_Y(Y_{(m)})}\right).$$

Therefore

$$E\left[\frac{-1}{m-1}\sum_{i=1}^{m-1} \log\left(\frac{F_Y(Y_{(i)})}{F_Y(Y_{(m)})}\right)\bigg| Y_{(m)}\right] = 1.$$

Given that $F_Y(y) = \Phi(y)y^\theta$, then

$$E\left[\frac{-\theta}{m-1}\sum_{i=1}^{m-1} \log\left(\frac{Y_{(i)}}{Y_{(m)}}\right)\bigg| Y_{(m)}\right] + E\left[\frac{-1}{m-1}\sum_{i=1}^{m-1} \log\left(\frac{\Phi(Y_{(i)})}{\Phi(Y_{(m)})}\right)\bigg| Y_{(m)}\right]$$

is equal to one. Rearranging, Equation 8.5 follows.

Now consider a Taylor series expansion of $1/\Theta_m$ about $1/\theta$, then take expectations to give

$$E\left[\frac{1}{\Theta_m}\bigg| Y_{(m)}\right] = \frac{1}{\theta} - \frac{E[(\Theta_m - \theta)|Y_{(m)}]}{\theta^2} + \frac{E[(\Theta_m - \theta)^2|Y_{(m)}]}{\theta^3} - \cdots.$$

Rearranging, and inserting Equation 8.5 gives Equation 8.6. Note also the relationship between the bias and the variance. □

8.4.5 Lemma

Given that all moments of the form $E[(\Theta_m - \theta)^k|Y_{(m)}]$ exist for all $k > 1$, then

$$E[(\Theta_m - \theta)^2|Y_{(m)}]$$

$$= \theta^4 E\left[\left(\frac{1}{\Theta_m} - \frac{1}{\theta}\right)^2\bigg| Y_{(m)}\right] + \frac{1}{\theta}E[(\Theta_m - \theta)^3|Y_{(m)}] - \cdots.$$

Proof. Form a Taylor series expansion of $1/\Theta_m$ about $1/\theta$, rearranging, and squaring both sides gives

$$\left(\frac{1}{\Theta_m} - \frac{1}{\theta}\right)^2 = \left(\frac{\Theta_m - \theta}{\theta^2}\right)^2 - \frac{(\Theta_m - \theta)^3}{\theta^5} + \cdots .$$

The result follows after rearranging and taking expectations.   □

8.4.6 Monotonicity of $\Phi(y)$

If $\Phi(y)$ is a monotonically decreasing function, then $\Phi(Y_{(i)}) \geq \Phi(Y_{(m)})$ since $Y_{(i)} \leq Y_{(m)}$. Therefore $\gamma_m > 0$. Assuming that the higher order terms in Equation 8.6 are sufficiently small, then Θ_m underestimates θ. This is consistent with Examples 7.2.2 and 7.2.3 where $\Phi(y)$ was monotonically decreasing, caused by the boundary effect.   □

8.4.7 Theorem (Mason, 1982)

Let $Y_1, Y_2, \cdots$ be i.i.d. random variables with common distribution $F_Y(y)$. Assume that $m = m(n) \to \infty$ and $m = o(n)$. Then Θ_m is a weakly consistent estimator of θ iff $F_Y(y)$ has a regularly varying lower tail with exponent θ (as in Definition 7.4.2).   □

Mikosch & Wang (1995) have further shown that if $m = n^\alpha$ for some $\alpha \in (0, 1)$, then Θ_m is strongly consistent with an asymptotic normal distribution.

8.4.8 Corollary

Given the conditions of Theorem 8.4.7, Θ_m is a weakly consistent estimator of θ iff $\Phi(y)$ is slowly varying at zero.

Proof. Follows from Theorem 8.4.7 and Proposition 7.4.5.   □

It follows from Corollary 8.4.8 that if $F_Y(y)$ is lacunary (Definition 7.4.6), then Θ_m cannot be a consistent estimator for θ. This gives Θ_m a serious deficiency when compared to the Grassberger-Procaccia method. The Takens estimator also suffers from the same deficiency. Smith (1992a) proposes a modification to the Takens estimator based on the beta-binomial distribution where the bias in the lacunar case appears to be reduced.

Results in this and the previous section have required the assumption that a sample of interpoint distances are independent. The justification for using such methods in general systems remains largely empirical. Some authors have suggested using a bootstrap procedure, where relatively few of the observed interpoint distances are randomly selected, as a means of enhancing the independence of selected pairs.

8.5 Bootstrap Estimation Procedure

The method of estimating of D_2 proposed by Grassberger & Procaccia (1983a) involved determining the powerlaw exponent of the empirical probability distribution of *all* interpoint distances. The Takens and Hill estimators are based on a random sample of independent interpoint distances, and both involve sums of the logarithms of distances that are sufficiently small. With fewer small interpoint distances than would be achieved with the Grassberger-Procaccia method, the Hill estimator Θ_m (or the Takens estimator Θ_ϵ) is quite variable for small m (or ϵ).

Mikosch & Wang (1995) advocated using a bootstrap sampling procedure to reduce the sample variance for small m (or ϵ). Such a bootstrap procedure also provides rough estimates of standard errors of the dimension estimates. Monte Carlo procedures were also advocated by Takens (1993). The bootstrap technique is discussed extensively by Efron & Tibshirani (1986, 1993).

8.5.1 Example - Cantor Measure

Consider the Cantor measure discussed in Example 7.3, in particular, where $p_0 = 0.5$. A random sample of $N = 100,000$, with probability distribution given by the Cantor measure, has been simulated by generating dyadic sequences of zeros and twos (of length 45 with the probability p_0 that a simulated digit is zero), and converting to a decimal number between zero and one.

Five sets of Hill estimates of D_2 have been calculated using the estimator given by Equation 8.4. Each set of estimations was based on random samples of $n = 100,000$ interpoint distances, and the Hill estimate has been calculated for various values of m. The value of m determines the largest order statistic of interpoint distances, $y_{(m)}$, to be used in the value of $\widehat{\theta}_m$. The value of m is a surrogate for the interpoint distance, and so a plot of m by $\widehat{\theta}_m$ would be quite similar to one of the interpoint distance by $\widehat{\theta}_m$. Usually one would select the range of m values so that $y_{(m)}$ spanned the interpoint distances where powerlaw behaviour was to be investigated. Evaluating more intermediate values of m would give greater resolution to the curve. One may alternatively plot $y_{(m)}$ by $\widehat{\theta}_m$ or $\log y_{(m)}$ by $\widehat{\theta}_m$. The latter is more appropriate if $F_Y(y)$ is lacunary, as in the current example, where the cyclical behaviour is periodic on a logarithmic scale.

Results are plotted in Figure 8.1. Note that there is considerable variability for small values of $y_{(m)}$. Also note that, because of the nature of the estimator, there is considerable autocorrelation in individual runs as m increases. It appears to be the case that individual runs often start too high or low, but after a number of cycles, converge closer to the expected value of Θ_m. One method to improve the estimates for smaller values of m would be to increase the number of sampled pairs n. The value of n would be determined so that values of $y_{(m)} \ll 3^{-11}$ are selected, and hence once $y_{(m)} > 3^{-11}$ the estimates may be more stable. This line of attack is difficult for computational reasons. In order to sample a few more

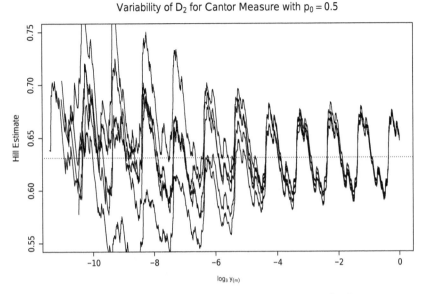

Figure 8.1 *Estimates of D_2 for the Cantor measure with $p_0 = 0.5$ for five separate sequences ($n = 100{,}000$, $N = 100{,}000$). The dotted line represents D_2.*

lacunary cycles, the number of pairs n needs to be at least doubled. One quickly starts to reach the limit of what is possible on many computers.

Another method is to repeat the estimation procedure many times, as in Figure 8.1, and calculate the average of the dimension estimate for each value of m. This is just the bootstrap method which has the added advantage that rough standard errors can be calculated for the dimension estimates for each value of m.   □

Our bootstrap procedure for estimating D_q ($q = 2, 3, \cdots$) differs from that described by Mikosch & Wang (1995). The reasons for this difference are highlighted in the following example.

8.5.2 Example - Exponential Distribution

The derivation of both the Takens and Hill estimators are based on an assumption that the interpoint distances have a powerlaw distribution. This assumption is not completely correct, particularly when there is lacunary like behaviour. However, even in this particular situation, a powerlaw distribution is an extremely good first order approximation; see for example, Figure 7.4. If the distribution was exactly powerlaw, then the line would be straight with slope given by the powerlaw exponent. The oscillatory behaviour is a manifestation of lacunarity. The logarithm of a random variable with a powerlaw distribution has an exponential distribution. In this example, we briefly consider this particular case.

Assume that Z is an exponential random variable with parameter λ. Then $E[Z] = \lambda^{-1}$ and $\text{Var}(Z) = \lambda^{-2}$. Consider k bootstrap sample estimates of the mean, each of sample size m. We are particularly interested in the case where m is quite small (sometimes < 10) but $k \approx 100$. Denote the i.i.d. estimators of the mean as $\overline{Z}_1, \cdots, \overline{Z}_k$. Now consider two estimators of λ,

$$\Lambda_1 = \frac{1}{k} \left(\frac{1}{\overline{Z}_1} + \cdots + \frac{1}{\overline{Z}_k} \right)$$

and

$$\Lambda_2 = \frac{k}{\overline{Z}_1 + \cdots + \overline{Z}_k} = \frac{1}{\overline{Z}^*},$$

where $\overline{Z}^*$ is the mean of all $m \times k$ sampled i.i.d. random variables. This would be the maximum likelihood estimator if all samples were collectively considered as one.

Since

$$E\left[\frac{1}{\overline{Z}_1}\right] = \frac{m}{m-1}\lambda \quad \text{and} \quad \text{Var}\left(\frac{1}{\overline{Z}_1}\right) = \left(\frac{m}{m-1}\right)^2 \frac{\lambda^2}{m-2},$$

then

$$E[\Lambda_1] = \frac{m}{m-1}\lambda \quad \text{and} \quad E[\Lambda_2] = \frac{mk}{mk-1}\lambda.$$

Similarly

$$\text{Var}(\Lambda_1) = \frac{1}{k}\left(\frac{m}{m-1}\right)^2 \frac{\lambda^2}{m-2} \quad \text{and} \quad \text{Var}(\Lambda_2) = \left(\frac{mk}{mk-1}\right)^2 \frac{\lambda^2}{mk-2}.$$

Hence it can be seen that Λ_2 not only has smaller bias, but also smaller variance. Typically, $k = 100$ and m could be as small as 5. In this case $E[\Lambda_1] = 5\lambda/4$, $E[\Lambda_2] = 500\lambda/499$, and $\text{Var}(\Lambda_1) \approx 2.5 \times \text{Var}(\Lambda_2)$.   □

8.5.3 Bootstrapped Hill Estimate

We initially set up a sampling framework that will be exploited more fully in Chapter 9. Let $x_1, \cdots, x_N$ denote the observed sequence of N d-dimensional vector observations. Also let $\mathcal{A}_0 = \{x_1, \cdots, x_N\}$ and $\mathcal{A}_\epsilon$ be a set that excludes some observations close to the boundaries, hence $\mathcal{A}_\epsilon \subseteq \mathcal{A}_0$. The number of observations excluded is determined by the value of ϵ, where $0 \leq \epsilon < \frac{1}{2}$.

Each bootstrap sample uses the following procedure. Let the index i represent the current bootstrap number. There are n sampled interpoint distances within each bootstrap sample; i.e., y_s where $s = 1, \cdots, n$.

1. For each s, sample x_{s_q} from $\mathcal{A}_\epsilon$ and $x_{s_1}, x_{s_2}, \cdots, x_{s_{q-1}}$ from $\mathcal{A}_0$. Then let

$$y_s = \max \left\{ \left\| x_{s_1} - x_{s_q} \right\|, \left\| x_{s_2} - x_{s_q} \right\|, \cdots, \left\| x_{s_{q-1}} - x_{s_q} \right\| \right\}.$$

2. Sort the differences to form the order statistics, where

$$y_{(1)} < y_{(2)} < \cdots < y_{(m)} < \cdots < y_{(n)}.$$

3. For various values of m, calculate

$$z_{i,m} = \frac{-1}{m-1} \sum_{j=1}^{m-1} \log \left(\frac{y_{(j)}}{y_{(m)}} \right). \tag{8.7}$$

The above steps are repeated k times, where k is the number of bootstrap samples taken. Then the bootstrapped Hill estimate is calculated as

$$\widehat{\theta}_m = \frac{k}{\sum_{i=1}^{k} z_{i,m}}. \tag{8.8}$$

Let $\bar{y}_{(m)}$ denote the average of the mth order statistic over all k bootstrap samples. Values of m are selected so that the interpoint distances $\bar{y}_{(m)}$ span the range where powerlaw behaviour of $F_Y(y)$ is to be analysed. Greater numbers of intermediate points are calculated to give the curve greater resolution. Possible plots are m by $\widehat{\theta}_m$, $\bar{y}_{(m)}$ by $\widehat{\theta}_m$, and $\log_b \bar{y}_{(m)}$ by $\widehat{\theta}_m$ where b is an appropriate base.

In the Grassberger-Procaccia method (§8.2) one plots $\log y$ by sample values of $\log \widehat{F}_Y^{GP}(y, N)$ and calculates the slope of the line to give $\widehat{\theta}$. In the above Hill method, $\widehat{\theta}$ is read directly off the vertical axis. In the lacunary case, determining where one 'draws the line' is not easy. The plot of the Hill estimates is essentially a scaled version of $\log \Phi(y)$, roughly centred (but not quite!) about θ. These difficulties will be discussed further in §8.6.

8.5.4 Standard Errors

The bootstrap procedure involves two levels of sampling. The initial sample is taken from $\mathcal{X}$, say $x_1, \cdots, x_N$. Then an empirical distribution is used, consisting of the points $x_i \in \mathcal{X}$, for $i = 1, \cdots, N$, each with probability $1/N$. The second level involves the bootstrap procedure, which involves repeated samplings from the empirical distribution. Standard errors calculated below are therefore conditional on the given empirical distribution.

Let $z_{i,m}$ be an observed value of the random variable $Z_{i,m}$, $i = 1, \cdots, k$, where k is the number of bootstraps. The random variables $Z_{1,m}, \cdots, Z_{k,m}$ are assumed to be i.i.d. with the same distribution as $1/\Theta_m$. Denote the bootstrap estimator (after k iterations) of θ as $\Theta_m^{(k)}$, where

$$\Theta_m^{(k)} = \frac{k}{\sum_{i=1}^{k} Z_{i,m}},$$

and let $\theta_m = \mathrm{E}\big[\Theta_m^{(k)}\big]$. Given an estimate of $\mathrm{Var}(Z_{1,m})$, which is an output of the bootstrap method, we want to derive an estimate of $\mathrm{Var}(\Theta^{(k)})$.

Lemma 8.4.5 also holds by substituting $\Theta_m^{(k)}$ and θ_m in place of Θ_m and θ respectively to give

$$
\mathrm{E}\left[\left(\Theta_m^{(k)} - \theta_m\right)^2 \middle| Y_{(m)}\right]
$$

$$
= \theta_m^4 \mathrm{E}\left[\left(\frac{1}{\Theta_m^{(k)}} - \frac{1}{\theta_m}\right)^2 \middle| Y_{(m)}\right] + \frac{1}{\theta_m}\mathrm{E}\left[\left(\Theta_m^{(k)} - \theta\right)^3 \middle| Y_{(m)}\right] - \cdots .
$$

The second term on the right-hand side can also be expanded with a Taylor series expansion, with a leading term involving $\mathrm{E}\left[\left(1/\Theta_m^{(k)} - 1/\theta_m\right)^3\right]$. Therefore, assuming independence of the bootstrap samples,

$$
\mathrm{E}\left[\left(\Theta_m^{(k)} - \theta_m\right)^2 \middle| Y_{(m)}\right] = \theta_m^4 \mathrm{Var}\left(\frac{Z_{1,m} + \cdots + Z_{k,m}}{k}\right) + o(k^{-2})
$$

$$
= \frac{\theta_m^4}{k}\mathrm{Var}(Z_{1,m}) + o(k^{-2}).
$$

An estimate of $\mathrm{Var}(Z_{1,m})$ is

$$
S_{Z_{1,m}}^2 = \frac{1}{k-1}\sum_{i=1}^{k} z_{i,m}^2 - \frac{k}{k-1}\left[\frac{1}{k}\sum_{i=1}^{k} z_{i,m}\right]^2, \tag{8.9}
$$

hence

$$
S_{\Theta_m^{(k)}}^2 = \frac{\widehat{\theta}_m^4}{k}S_{Z_{1,m}}^2 \tag{8.10}
$$

provides an estimate of $\mathrm{Var}(\Theta_m^{(k)})$, where $\widehat{\theta}_m$ and $S_{Z_{1,m}}$ are given by Equations 8.8 and 8.9 respectively. $\qquad\square$

8.6 Discussion and Examples

As already stated in Corollary 8.4.8, the Hill estimator will not be consistent when $F_Y(y)$ is lacunary. In this section, we initially consider two examples: the uniform distribution, where the conditions of Corollary 8.4.8 are satisfied; and the Cantor measure where they are not. We will attempt to identify the source of the bias in the Hill estimator. We will also contrast the lack of consistency of the Hill estimator with the Grassberger-Procaccia method.

8.6.1 Example - Uniform Distribution

A sample from the uniform distribution on $[0, 1]$ has been simulated ($N = 10{,}000$) and Hill estimates of D_2, using Equation 8.8, for various values of m have been plotted in Figure 8.2. Each of the $k = 100$ bootstrap samples consisted of $n = 10{,}000$ interpoint distances. The horizontal axis, $\log_{10}\bar{y}_{(m)}$, is the average over all $k = 100$ bootstraps of the mth order statistic $y_{(m)}$. Dimension estimates are often

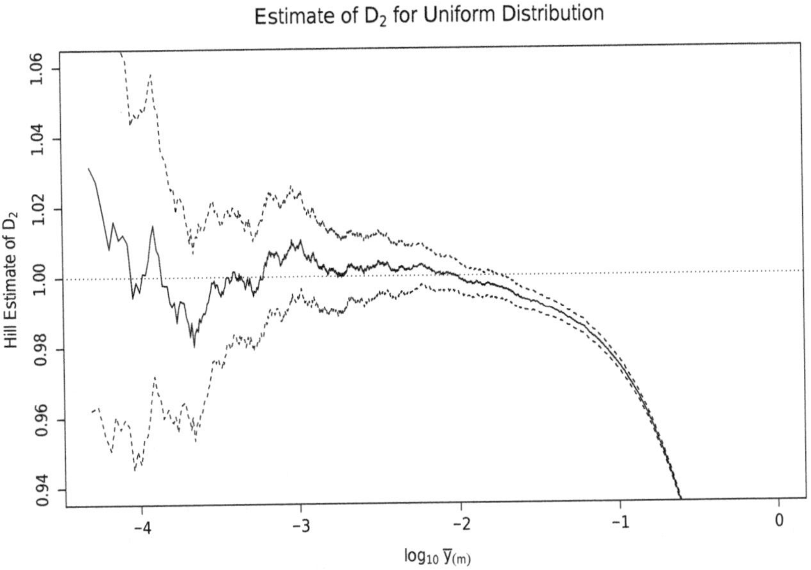

Estimate of D_2 for Uniform Distribution

Figure 8.2 *Estimate of D_2 for the uniform distribution ($k = 100$, $n = 100{,}000$, $N = 100{,}000$). The upper and lower dashed lines mark the interval width given by two standard errors.*

plotted against the logarithm of the interpoint distance, particularly when there is lacunary behaviour. Intervals of twice the standard error have been calculated using Equation 8.10 and overlaid.

From §7.2.3, $F_Y(y) = y^d(2 - y)^d$, and hence $\Phi(y) = (2 - y)^d$ is slowly varying at zero, here $d = 1$. As implied by Corollary 8.4.8, the estimates are consistent, and are asymptotically unbiased. However, these estimates suffer from the boundary effect discussed in §7.2 and to be discussed further in §9.2. In order to get a reasonably accurate dimension estimate, one must sample sufficiently small interpoint distances so that the function $F_Y(y)$ is dominated by the powerlaw component, with the $\Phi(y)$ component having little effect.

In higher dimensions, the likelihood of sampling small interpoint distances decreases. See for example the dimension estimates of D_2 in Figure 10.13 of what is essentially white noise in $\mathbb{R}^d$ for $d = 1, \cdots, 5$. Powerlaw behaviour is evident for $d = 1$, but as d increases, the boundary effect becomes more of a problem with $\Phi(y)$ becoming more dominant. This will be discussed in §9.2 and is not peculiar to the Hill method of estimation.   □

Now consider an example whose characteristics are opposite in respect of what was discussed in the preceding example. The probability distribution, $F_Y(y)$, derived from the Cantor measure is lacunary and therefore, according to Corollary 8.4.8, the Hill estimator is not consistent. However, in this example, a form of powerlaw behaviour extends over the entire unit interval (see Definition 7.4.6), and it is not affected by the boundary effect in the same manner as in the previous example.

8.6.2 Example - Cantor Measure

A random sample of $N = 100,000$, with probability distribution given by the Cantor measure, has been simulated as in Example 8.5.1. Estimates of $\hat{\theta}_m$ (Equation 8.8), are plotted in Figures 8.3 and 8.4 for $p_0 = 0.5$ and 0.2 respectively. Estimates are plotted by both m and $\log_3 \bar{y}_{(m)}$, where $\bar{y}_{(m)}$ is the average over all $k = 100$ bootstraps of the mth order statistic $y_{(m)}$, and $n = 100,000$ interpoint distances were sampled. The dotted line in Figures 8.3 and 8.4 represents the true value of D_2, which can be calculated by appealing to Theorem 5.5.4, then using Equations 2.3 and 3.8. It follows from Corollary 8.4.8 that the Hill estimator is not consistent in this example, clearly seen in both Figures 8.3 and 8.4.

Note that the standard error, given by Equation 8.10, increases as $\bar{y}_{(m)} \to 0$ (bottom graph). The plot of the standard errors is roughly linear (on the log-log plot), hence a powerlaw increase as $\bar{y}_{(m)} \to 0$. It also contains lacunary like behaviour, which is more evident in Figure 8.4. When one samples a fixed number of interpoint distances, one selects smaller values when $p_0 = 0.2$, i.e., when the measure is not distributed uniformly on the support. Consequently, one can observe more lacunary cycles in this situation.

One question that arises from these plots is where one draws the line representing the estimate of D_q. A first approximation may be to place the line representing D_q in the 'middle', midway between the peaks and troughs of the lacunary cycles. However, it will be seen later that in the case of D_5 in particular, this is obviously not in the 'middle' of the lacunary cycles. In fact, in all of these plots, using the 'middle' as the estimate of D_q would tend to induce a positive bias.   □

8.6.3 Where to 'Draw' the Line

In order to decide on where one draws the line, particularly when there is lacunary behaviour, we return to Equation 8.6, which gave the expected value of the Hill estimator as

$$E\left[\Theta_m | Y_{(m)} = y\right] = \theta[1 - \gamma_m(y)] + \frac{E\left[(\Theta_m - \theta)^2 | Y_{(m)} = y\right]}{\theta} - \cdots , \quad (8.11)$$

where

$$\gamma_m(y) = E\left[\frac{1}{m-1} \sum_{i=1}^{m-1} \log\left(\frac{\Phi(Y_{(i)})}{\Phi(Y_{(m)})} \right) \middle| Y_{(m)} = y \right].$$

Now assume that as m becomes sufficiently large

$$\gamma_m(y) \approx E[\log \Phi(Y) | Y < y] - \log \Phi(y),$$

Estimate of D_2 for Cantor Measure with $p_0 = 0.5$

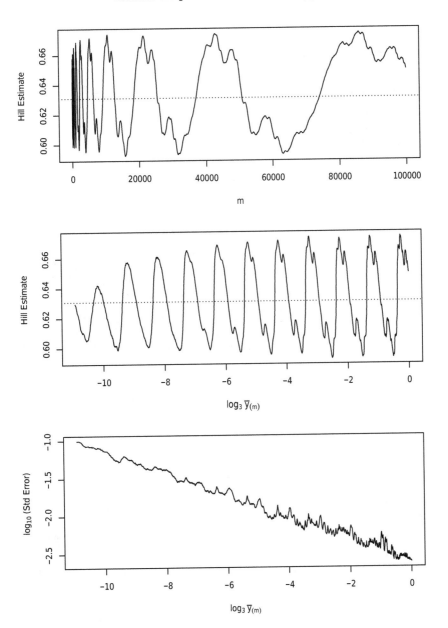

Figure 8.3 *Estimate of D_2 for the Cantor measure with $p_0 = 0.5$ ($k = 100$, $n = 100,000$, $N = 100,000$). The dotted line represents D_2.*

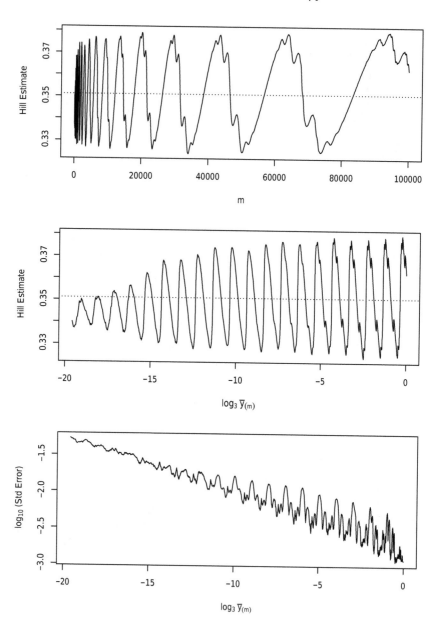

Figure 8.4 *Estimate of D_2 for the Cantor measure with $p_0 = 0.2$ ($k = 100$, $n = 100,000$, $N = 100,000$). The dotted line represents D_2.*

and that higher order terms in Equation 8.11 are negligible; then

$$\begin{aligned} E\left[\Theta_m | Y_{(m)} = y\right] &\approx \theta[1 - \gamma_m(y)] \\ &\approx \theta + \theta \log \Phi(y) - \theta \, E[\log \Phi(Y) | Y < y]. \end{aligned} \quad (8.12)$$

It may be tempting to argue that if $F_Y(y)$ is lacunary, then $E[\log \Phi(Y)|Y < y]$ will contain an infinite number of cycles as $y \to 0$ for any value of y, and hence should have reached a limiting constant. Unfortunately, this is not the case. The expectation is being taken with respect to the probability distribution of Y, which by assumption is powerlaw. The increasing likelihood of larger interpoint distances causes this expectation to also have similar periodic behaviour as $\log \Phi(y)$, though with a much smaller amplitude. Therefore, we would expect the dimension estimates to be essentially an estimate of the $\log \Phi(y)$ function, 'centred' at θ, but where the centering moves slightly in a periodic fashion. This can be seen from Figure 8.3 which, at least visually, appears to be a scaled and shifted version of $\log \Phi(y)$ as plotted in Figure 7.5.

An indication of where one draws the line is given by considering conditional expectations of Equation 8.12, in particular

$$\begin{aligned} E\left[E[\Theta_m | Y_{(m)} = Y] \,\middle|\, Y < y\right] \\ \approx \theta + \theta \, E[\log \Phi(Y) | Y < y] - \theta \, E[E[\log \Phi(Y) | Y < Y'] | Y' < y]. \end{aligned} \quad (8.13)$$

This is equivalent in the sample context, of Figures 8.3 and 8.4, to taking running partial means over the dimension estimates; effectively smoothing the lacunary cycles. It will be seen in the following example that the difference between the two terms inducing the bias in Equation 8.12 is greatly reduced in Equation 8.13. Repeated conditional expectations (i.e., smoothing), assuming that the dimension estimates are sufficiently stable, further reduces the effect of the bias terms.

8.6.4 Cantor Example Continued

Further estimates of $D_2, \cdots, D_5$ for the Cantor measure can be found in Figures 8.5 and 8.6. In both situations sample sequences of length $N = 100{,}000$ have been generated as in Figures 8.3 and 8.4. The $k = 100$ bootstrap samples each consisted of $n = 100{,}000$ interpoint distances of order q.

The lines with the greatest amplitudes are the unsmoothed bootstrap estimates given by Equation 8.8. The line with the amplitudes of intermediate size represents the partial means (i.e., one level of smoothing) suggested by Equation 8.13. The line with the smallest amplitudes represents the Hill estimates after two levels of smoothing. Further smoothing reduces further the oscillatory behaviour. It should be noted that the smoothed estimates do not become stable until $\overline{y}_{(m)}$ becomes reasonably large. This instability is caused by the instability of the raw Hill estimates. In each level of smoothing, this instability seems to extend further to the right into higher values of $\overline{y}_{(m)}$. Thus there is a practical limitation on the number of smoothings that can be performed.

Dimension Estimates for Cantor Measure with $p_0 = 0.5$

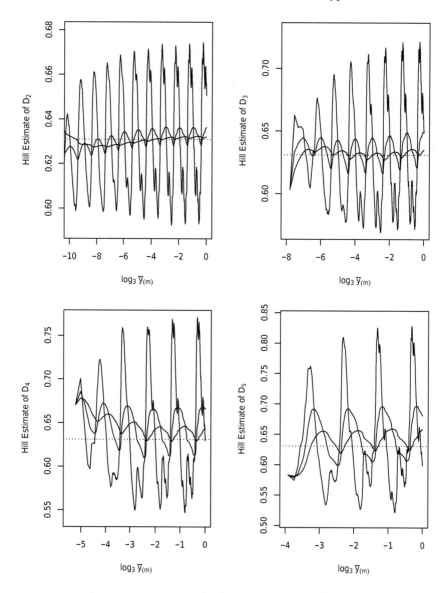

Figure 8.5 *Estimates of $D_2, \cdots, D_5$ for the Cantor measure with $p_0 = 0.5$ ($k = 100$, $n = 100{,}000$, $N = 100{,}000$). Lines with the greatest amplitudes are the unsmoothed estimates given by Equation 8.8. Lines with the amplitudes of intermediate size represent one level of smoothing, and lines with the smallest amplitudes represent two levels of smoothing. The horizontal dotted line represents the known Rényi dimension D_q.*

Dimension Estimates for Cantor Measure with $p_0 = 0.2$

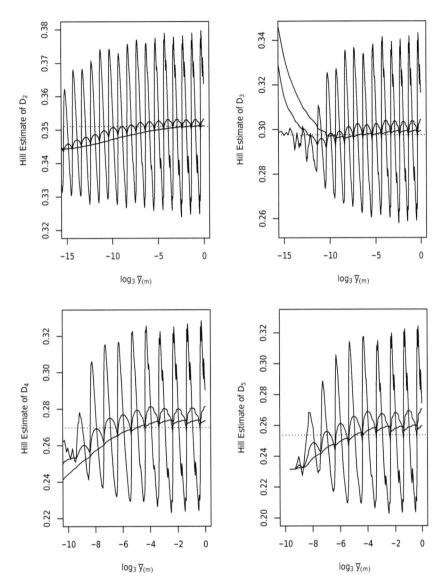

Figure 8.6 *Estimates of* $D_2, \cdots, D_5$ *for the Cantor measure with* $p_0 = 0.2$ $(k = 100,\ n = 100{,}000,\ N = 100{,}000)$. *Lines with the greatest amplitudes are the unsmoothed estimates given by Equation 8.8. Lines with the amplitudes of intermediate size represent one level of smoothing, and lines with the smallest amplitudes represent two levels of smoothing. The horizontal dotted line represents the known Rényi dimension* D_q.

In both figures, for larger interpoint distances, the smoothed lines oscillate quite closely to the true value of the Rényi dimension. However, note that there is still a positive bias in the smoothed estimates.   □

8.6.5 Bias Reduction

In the situation where there is a clearly defined lacunary cycle, an initial estimate of the Rényi dimension can be calculated as the mean of the smoothed estimates in the last lacunary cycle. As can be seen from Figures 8.5 and 8.6, this will produce estimates with a positive bias. These values have been tabulated as the 'uncorrected estimate' in Table 8.1.

The first excluded term in Equation 8.11 involves $E\left[(\Theta_m - \theta)^2 | Y_{(m)}\right]$. Note that

$$E\left[(\Theta_m - \theta)^2\right] = E\left[(\Theta_m - E[\Theta_m])^2\right] + (E[\Theta_m] - \theta)^2 \,,$$

where each expectation is conditional on $Y_{(m)}$. An estimate of $\text{Var}(\Theta_m)$ is $S^2_{\Theta_m}$, the bootstrap variance given by Equation 8.10. However, this term is relatively small compared to that of $(E[\Theta_m] - \theta)^2$ and so we ignore it at this point. This suggests that a better approximation than that in Equation 8.12 is

$$E\left[\Theta_m | Y_{(m)} = y\right] - \frac{(E[\Theta_m] - \theta)^2}{\theta}$$

$$\approx \quad \theta + \theta \log \Phi(y) - \theta \, E[\log \Phi(Y) | Y < y] \,.$$

Hence a 'corrected estimate' of D_q can be calculated by subtracting the variance term from the raw Hill estimates (given by Equation 8.8), as on the left-hand side of the above equation, but using the 'uncorrected estimate' in place of θ. Then

q	Cantor Measure with $p_0 = 0.5$			Cantor Measure with $p_0 = 0.2$		
	Actual Value	Uncorrected Estimate	Corrected Estimate	Actual Value	Uncorrected Estimate	Corrected Estimate
2	0.6309	0.6320	0.6310	0.3510	0.3512	0.3503
3	0.6309	0.6330	0.6294	0.2976	0.2992	0.2969
4	0.6309	0.6365	0.6295	0.2696	0.2730	0.2689
5	0.6309	0.6406	0.6290	0.2537	0.2584	0.2527
6	0.6309	0.6515	0.6349	0.2437	0.2504	0.2428
7	0.6309	0.6670	0.6451	0.2370	0.2446	0.2353
8	0.6309	0.6886	0.6618	0.2321	0.2413	0.2301

Table 8.1 *Estimates of the Rényi dimensions, D_q, for the Cantor measure. In both the corrected and uncorrected estimates, two passes of smoothing were performed as shown in Figures 8.5 and 8.6.*

one would smooth as in Equation 8.13. In Table 8.1, the 'corrected estimate' is the average of the smoothed and corrected Hill estimates in the last lacunary cycle.

□

In situations where the non-powerlaw component $\Phi(y)$ not only contains a lacunary like component, but also a boundary effect, then the situation is more complicated. Examples of this situation will be given in §10.2.

8.6.6 Discussion

We have discussed essentially two methods of estimating the Rényi dimensions, one based on maximum likelihood (Takens and Hill) and one on the slope of a log-log plot (Grassberger-Procaccia). Both methods have their problems but also their relative advantages. We briefly summarise some of the respective advantages and disadvantages.

Both methods suffer from a possible boundary effect which will be discussed further in Chapter 9. The boundary effect does not always occur, though when it does, it causes a deficit of larger interpoint distances whose size is roughly comparable to the region width. That is, the powerlaw relationship does not hold for large interpoint distances that are comparable to the region width. Both methods also suffer from greater variability at smaller interpoint distances caused by smaller sample sizes.

The Grassberger-Procaccia method of estimating the correlation exponents involves plotting $\log y$ by $\log \widehat{F}_Y^{GP}(y, N)$. An estimate of the correlation exponent is given by the estimated slope of the line. Often some form of least squares regression is used to estimate the slope. Unfortunately, there are some arbitrary choices that need to be made. An interval needs to be selected where the line is sufficiently straight that is not affected by the boundary effect at the upper end, and possibly extreme variability and lack of data at the lower end. Another problem is that one is fitting a line to an effectively cumulative sum of counts. Thus points entering into the regression are obviously not independent, and have different variances caused by different sample sizes.

The Takens and Hill estimators are based on the theory of maximum likelihood and therefore carry some of the nice properties associated with that theory. Another advantage is that these estimators are more explicitly defined and, therefore, lend themselves more easily to having their properties analysed. However, while one can derive estimates of the required Rényi dimensions over a range of interpoint distances, and evaluate the bias and variance at each of these points, they still suffer from the boundary effect and variability at the lower end as does the Grassberger-Procaccia method. An advantage of both the Takens and Hill estimation methods over the Grassberger-Procaccia method is that the resultant plot is directly that of the dimension estimates. In the Grassberger-Procaccia method, one needs to calculate the slope or local derivatives with the inherent difficulties that this causes. The plots based on the Hill method also tend to accentuate any lacunary behaviour which, in some ways, is as interesting as the fractal dimension.

It not only tells us that the measure is supported on a self-similar like set, but also tells us the scaling factor.

An advantage of the Grassberger-Procaccia method is that it is consistent, while the Takens and Hill methods are not. Assume that we had an extremely large sample size, and that a plot of $\log y$ by $\log \widehat{F}_Y^{GP}(y, N)$ had been done as prescribed by this method. In the case of the Cantor measure, the plot would look like that in Figure 7.4. As we continue to increase the sample size, we would sample smaller and smaller interpoint distances, and the plot, with further lacunary cycles would be extended to the left. As the line is extended further to the left, with an increasing number of lacunary cycles, it can be seen that the possible slope of the fitted line gets sandwiched closer and closer to the true slope. Alternatively, it can be seen from Figures 8.3 and 8.4, that as the sample size increases in the Hill method, further lacunary cycles are added to the plots for smaller values of $\bar{y}_{(m)}$, though it is no clearer where one should 'draw the line'.

This may not be such a problem though. It appears from our calculations in this section, using the Cantor measure, that reasonably accurate dimension estimates can be calculated using the Hill method. No ad hoc decisions were required to determine the 'straight' part of the line. It appears from our numerical studies that if the measure under analysis has 'gaps' like the Cantor measure, then conditional expectations as in Equation 8.13 will reduce the amplitude of the Hill estimates closer to the value of θ, apart from the higher order bias terms as in Equation 8.11. The 'gaps' appear to ensure that $\Phi(y)$ does not contain an overall boundary component (see Example 10.2.1).

CHAPTER 9

Extrinsic Sources of Bias

9.1 Introduction

In Chapter 7, it was shown that if the correlation exponent $\theta(q)$ exists, then the probability function $F_Y(y)$ could be decomposed as $\Phi(y)y^{\theta(q)}$, where $\Phi(y)$ denotes the non-powerlaw part. The non-powerlaw behaviour induced by the function $\Phi(y)$ was referred to as an intrinsic form of bias. In Chapter 8, methods to estimate

$$D_q = \frac{\theta(q)}{q-1}$$

for $q = 2, 3, \cdots$, were investigated.

Because of experimental limitations or deficiencies, the empirical probability distribution of sampled interpoint distances can have quite different characteristics to that of $F_Y(y)$ given by Equation 7.6. We are not referring to those sources of bias that are related to the particular estimator used as in Chapter 8. The extrinsic sources of bias referred to here are analogous to non-sampling error in sample survey methodology, are an inherent part of the data, but not characteristic of the underlying process (i.e., not intrinsic bias).

In many 'real' situations, such data deficiencies can be minimised but probably not eliminated. Many datasets consist of observations that have been collected for many years, long before analyses often performed on them today were ever thought about. Take, for example, catalogues of earthquake locations. The earthquake process is part of a large system of events worldwide, but occurring mostly on or near tectonic plate boundaries. Detection of all events above a given magnitude requires the event to be sufficiently close to an active seismic network. Effectively the 'boundary' is determined by the efficacy of the seismic network. Within the region where events can be detected, the accuracy of the determined event location may also vary considerably. However, seismic networks are being continually upgraded, thus boundaries and location errors (noise) are continually changing.

The effect of extrinsic sources of bias on dimension estimates are quite difficult to quantify analytically. There are three cases of extrinsic bias we investigate in this chapter: imposed boundary effect, rounding effect and the effect of noise in the data. We use the Hill estimator with a bootstrap sampling procedure (§8.5) to demonstrate the above effects on dimension estimates where data have been simulated.

9.2 Imposed Boundary Effect

The methods described in Chapter 8 for estimating the Rényi dimensions are based on interpoint distances. When an artificial boundary is imposed, and only data within the region defined by that boundary are collected, excluding data from outside of the region, a boundary effect occurs.

Take, for example, the interpoint distance when $q = 2$. In order to achieve an interpoint distance that is comparable to the width of the region, both points must be sampled from areas that are close to opposite boundaries. Conversely, there are many more possibilities for sampling a smaller interpoint distance. Consequently, there is a deficit of larger interpoint distances relative to the situation where there are no imposed boundaries.

In this section, the effect of observing a smaller region than that in which the process is operating is described; that is, the effect of imposing a boundary within which data will be sampled. These imposed boundaries are conceptually different to those boundaries that are intrinsic to the process, as described in Chapter 7, but they have much the same effect. A partial correction to the boundary effect is given.

9.2.1 Boundary Effect Correction

One method of adjusting for the boundary effect is to subsample from a subregion of the observation region. Let $\mathcal{A}_0$ be a set containing all observations within the observation region, i.e.,

$$\mathcal{A}_0 = \{x_j : j = 1, \cdots, N\},$$

where $x_j \in \mathbb{R}^d$. Let $\mathcal{A}_\epsilon$ be a smaller set of points, where $\mathcal{A}_\epsilon \subseteq \mathcal{A}_0$. Observations that are close to the boundaries of $\mathcal{A}_0$ are excluded from $\mathcal{A}_\epsilon$. $\mathcal{A}_\epsilon$ is defined as follows.

Let $x_j = (x_{j1}, \cdots, x_{jd})$. Denote the range of points in the uth dimension as

$$r_u = \max_j(x_{ju}) - \min_j(x_{ju}) \qquad u = 1, \cdots, d.$$

Then define the reduced set of points $\mathcal{A}_\epsilon$, for $0 \leq \epsilon < \frac{1}{2}$, as

$$\mathcal{A}_\epsilon = \left\{ x_j : x_j \in \prod_{u=1}^d \left[\min_j(x_{ju}) + r_u\epsilon, \max_j(x_{ju}) - r_u\epsilon \right], \quad j = 1, \cdots, N \right\}.$$

$$(9.1)$$

The sampling procedure then proceeds as in §8.5. Let the sth ($s = 1, \cdots, n$) interpoint distance of order q be

$$y_s = \max \left\{ \left\| x_{s_1} - x_{s_q} \right\|, \left\| x_{s_2} - x_{s_q} \right\|, \cdots, \left\| x_{s_{q-1}} - x_{s_q} \right\| \right\}.$$

By sampling x_{s_q} from $\mathcal{A}_\epsilon$ and $x_{s_1}, x_{s_2}, \cdots, x_{s_{q-1}}$ from $\mathcal{A}_0$ the boundary effect is not very noticeable for interpoint distances that are less than the width of the inner region $\mathcal{A}_\epsilon$. $\qquad\qquad\square$

9.2.2 Example - Uniform Distribution

Figure 9.1 shows dimension estimates of D_2 and D_4 for uniformly distributed points on the unit interval $[0, 1]$; where $N = 10^4$ points were simulated, and $k = 100$ bootstrap samples were selected each containing $n = 10^4$ interpoint distances. Dimension estimates were calculated using Equation 8.8 for both $\epsilon = 0$ and $\epsilon = 0.3$. Results are plotted by $\overline{y}_{(m)}$ on the horizontal scale. It is the average over all k bootstraps of the mth order statistic $y_{(m)}$.

Using a value of $\epsilon = 0.3$ makes the width of the inner region 0.4. It can be seen in Figure 9.1 that the dimension estimates are one until the interpoint distances get close to the width of the inner region, then drop even more sharply than the non-corrected lines.   □

Consider N uniformly distributed points in a d-dimensional unit cube. Then the expected number of points in $\mathcal{A}_\epsilon$, given by Equation 9.1, is $N(1 - 2\epsilon)^d$. Consequently, as d increases, the number of expected points in the restricted region decreases at a powerlaw rate. Hence the method is somewhat wasteful of data.

Another problem with the above method is that sampling boundaries are often not clearly defined and straightforward as in the above example. The boundary of a seismic network within which earthquake events, greater than some prescribed magnitude, will be detected with a high probability, is often extremely irregular. It

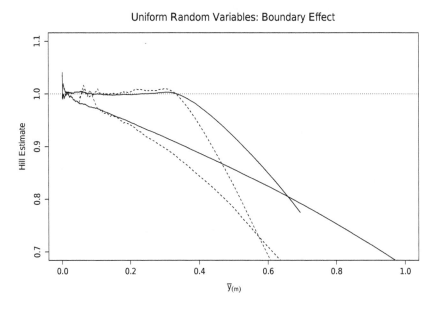

Figure 9.1 *Boundary effect correction by excluding a fraction of $\epsilon = 0.3$. The solid lines are estimates of D_2, the upper one having the boundary correction. Similarly, the dashed lines are estimates of D_4 ($k = 100$, $n = 10,000$, $N = 10,000$).*

will not only depend on the density of the network, but on the geological characteristics of the area which will also affect whether certain events can be detected.

Another method that is less wasteful of data and applicable to the Grassberger-
Procaccia algorithm (§8.2) is given by Stoyan & Stoyan (1994, §5.5) for $q = 2$.
In the Grassberger-Procaccia procedure, counts are made of the number of points
within a distance δ of a given point. The number of points found will be too
small if the given point is close to an imposed boundary. Stoyan & Stoyan (1994)
suggest scaling the counts by the volume of the sphere about the given point
divided by the volume of that sphere that intersects with the region under analysis.

Dvořák & Klaschka (1990) suggest another correction based on the uniform
distribution. The correlation integral for points in a d-dimensional unit cube is
$F_Y(y) = y^d(2 - y)^d$ for $q = 2$. They suggest modifying the Grassberger-
Procaccia (1983a) algorithm by calculating the slope of $\log F_Y(y)$ vs $\log[y(2 - y)]$, having already scaled the points into the unit cube.

9.2.3 Example - Bimodal Beta Distribution

The beta distribution provides an interesting counter example to the boundary
effect. When $\alpha = \beta = 1$, the beta distribution coincides with the uniform distribution. But when $\alpha = \beta$ are less than one, the distribution is bimodal with
increasing concentration of mass at either end of the interval $[0, 1]$ as α and β
decrease. Since the beta distribution has a continuous density, then $D_q = 1$ for
all values of $q \geq 1$; however if $\alpha = \beta$ is sufficiently small, $F_Y(y)$ for $q = 2$ is
approximately powerlaw with exponent 2α over almost the entire range of y. This
example will be discussed further in §10.3.2.   □

9.3 Rounding Effect

Most data are rounded if only to the extent of the number of digits stored to represent such a number in a database. Often a level of rounding occurs that equates
in some way to the accuracy of the observations. If data rounding is sufficiently
severe, there is a noticeable effect on dimension estimates. Data rounding can
have two main effects. There occurs a number of zero distances between sampled
pairs of points. This causes a problem with all estimators where logarithms of interpoint distances are required. Another behaviour is a modulating effect, which
should not be confused with the lacunary behaviour discussed in Chapter 7.

One possible method to handle the zero differences is to modify the Hill estimator by appealing to the truncated exponential distribution.

9.3.1 Truncated Exponential Approximation

Consider a sample of n interpoint distances $y_1, \cdots, y_n$ in $[0, 1]$ with probability distribution $F_Y(y) = y^\theta$. Assume that these observations have been subsequently rounded to the values $0, s, 2s, 3s, \cdots$, where $0 < s < 1$; and that h of the

rounded values equal zero, and $n - h$ are non-zero. Let Y_i' be the rounded value of the random variable Y_i. Assuming that the rounding has not had a too adverse effect on the distribution of the non-zero values, $W_i' = -\log Y_i'$ will have an approximate truncated exponential distribution. For notational convenience, let $y_1, \cdots, y_{n-h}$ denote the non-zero sample values and $y_{n-h}, \cdots, y_n$ denote the zero sample values.

Letting $t/2 = -\log(s/2)$, it follows that $\Pr\{Y_i' = 0\} = \Pr\{Y_i < s/2\} = \Pr\{W_i > t/2\} = \exp(-\theta t/2)$, thus using the truncated exponential distribution, the log-likelihood is

$$L(\theta; w_1, \cdots, w_n)$$

$$= \log\left[\binom{n}{h}(e^{-\theta t/2})^h(1 - e^{-\theta t/2})^{n-h}\prod_{i=1}^{n-h}\frac{\theta e^{-\theta w_i}}{1 - e^{-\theta t/2}}\right]$$

$$= \log\left[\binom{n}{h}e^{-\theta h t/2}\prod_{i=1}^{n-h}(\theta e^{-\theta w_i})\right].$$

The maximum likelihood estimate of θ is then given by

$$\widehat{\theta} = \frac{n - h}{ht/2 + \sum_{i=1}^{n-h} w_i}. \tag{9.2}$$

9.3.2 Modified Hill Estimator

Using Equation 9.2, the bootstrapping procedure of §8.5 is modified as follows. Equation 8.7 becomes

$$z_{i,m} = \frac{-1}{m - h}\left[h\log\left(\frac{s}{2y_{(m)}}\right) + \sum_{\substack{j=1 \\ y_{(j)}\neq 0}}^{m-1}\log\left(\frac{y_{(j)}}{y_{(m)}}\right)\right],$$

and Equation 8.8 remains

$$\widehat{\theta}_m = \frac{k}{\sum_{i=1}^{k} z_{i,m}},$$

where k is the number of bootstraps, and h is the number of sampled zero differences.

9.3.3 Further Problems

The previous scheme is not entirely correct, because in practice the rounding is implemented at an earlier stage, i.e., on the individual point positions X_i, where $i = 1, \cdots, N$. Let X_i be the original point position and X_i' be the corresponding rounded value. Assume that the random variable X_i' has been rounded to take possible discrete values $x_0, x_0 + s, x_0 + 2s, \cdots$, where $s > 0$. Then let $Y' = |X'_1 - X'_2|$ and $Y = |X_1 - X_2|$. Note the following.

1. $Y' = 0$ iff both X'_1 and X'_2 originate from the *same* rounded state. Further, $Y' = 0 \Rightarrow Y \in (0, s)$, however $Y' = 0 \nLeftarrow Y \in (0, s)$.

2. $Y' = s$ iff X'_1 and X'_2 originate from *neighbouring* states. Further, $Y' = s \Rightarrow Y \in (0, 2s)$, however $Y' = s \nLeftarrow Y \in (0, 2s)$.

3. $Y' = 2s$ iff X'_1 and X'_2 originate from *alternate* states. Further, $Y' = 2s \Rightarrow Y \in (s, 3s)$, however $Y' = 2s \nLeftarrow Y \in (s, 3s)$.

4. Et cetera.

Hence the rounded differences represent *overlapping* intervals on the Y scale, and the first two states of Y' could *both* represent very small interpoint distances. Using $Y' = 0$ as the truncated tail in a truncated exponential distribution, with a truncation point of s, will give a deficit of zeros, as many values less than s could be contained in the state $Y' = s$. This would cause the estimate of θ to be too large.

By using the estimate given by Equation 9.2, one is using the smaller 'truncation' point of $s/2$ for $Y' = 0$. This partially compensates for those small differences included in $Y' = s$. By using such a truncation value, one can also treat non-zero values of Y' as representing non-overlapping intervals of equal length.

□

9.3.4 Example

Let X_1 and X_2 be i.i.d. random variables sampled from a continuous uniform distribution on $[0, 1]$. After rounding to p decimal places, X'_1 can have $10^p + 1$ possible discrete states, with probability function

$$\Pr\{X'_1 = x\} = \begin{cases} \dfrac{1}{2 \times 10^p} & x = 0, 1 \\[2ex] \dfrac{1}{10^p} & x \neq 0, 1. \end{cases}$$

Note that we do not require p to be an integer, only $10^p + 1$ must be a positive integer. Now let $Y' = |X'_1 - X'_2|$. Then it can be shown that

$$\Pr\{Y' = y\} = \begin{cases} \dfrac{2 \times 10^p - 1}{2 \times 10^{2p}} & y = 0 \\[2ex] \dfrac{2 - 2y}{10^p} & y = \dfrac{1}{10^p}, \dfrac{2}{10^p}, \cdots, \dfrac{10^p - 1}{10^p} \\[2ex] \dfrac{1}{2 \times 10^{2p}} & y = 1. \end{cases} \qquad (9.3)$$

As such, the distribution of Y' closely follows the triangular distribution of the continuous unrounded case when $1/10^p \leq y \leq (10^p - 1)/10^p$. At first there appears to be a considerable deficit of point differences for $y = 0, 1$. This is not the case, as $Y' = 0, 1$ represent intervals that are half the width of those represented by other states.

Therefore, if we think of $Y' = 0$ as representing those values of $Y \in (0, s/2)$, $Y' = s$ as representing those values of $Y \in (s/2, 3s/2)$, etc., then the probability distribution of Y' gives a good approximation to Y. In a sense, the distribution of Y' is not only a discretised version of Y, but also smoothed.   □

9.3.5 Example

Figure 9.2 shows the effect of rounding simulated uniform random variables. $N = 10{,}000$ scalar uniform random variables were simulated on $[0, 1]$. Each value was then rounded to p decimal places, where $\log_{10} p = 0.05$, i.e., each element has been rounded to the states $0, 0.05, 0.10, \cdots, 1.00$. The bootstrapped Hill estimates of D_2 are plotted (solid line) and the dashed line represents the dimension estimates when no rounding is done.

The modulating effect is caused by the severe rounding. This should not be confused with the lacunary effect discussed in §7.4. Note that the modulations in the case of rounding have constant period on the $\bar{y}_{(m)}$ scale, creating a saw tooth

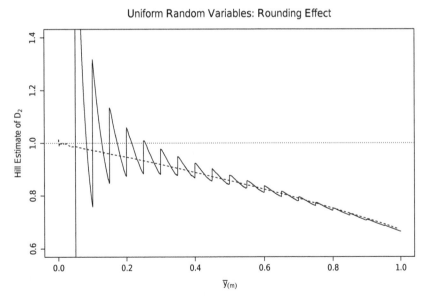

Figure 9.2 *Hill estimates* $(q = 2)$ *when sampled points are uniformly distributed on the unit interval but have been rounded to* $0, 0.05, 0.1, 0.15 \cdots, 1$. *The horizontal dotted line is for reference only, and the dashed line is the Hill estimates where no rounding has been done on the data* $(k = 100, \; n = 10{,}000, \; N = 10{,}000, \; \epsilon = 0)$.

effect, with decreasing amplitude as $\overline{y}_{(m)}$ increases. It is caused by the incrementing of $y_{(m)}$ in the denominator of Equation 8.4. When $y_{(m)} = 0.05$, all values of the order statistics are either 0 or 0.05. Therefore, when $y_{(m)}$ increases to 0.10, there is a considerable effect because for nearly all values of $k < m$, $y_{(k)}$ will be 0 or 0.05, hence the vertical jump. As $y_{(m)}$ increases, more values of $y_{(k)}$ will be 0.1, hence the line decreases until $y_{(m)} = 0.15$. The cycle then repeats itself.   □

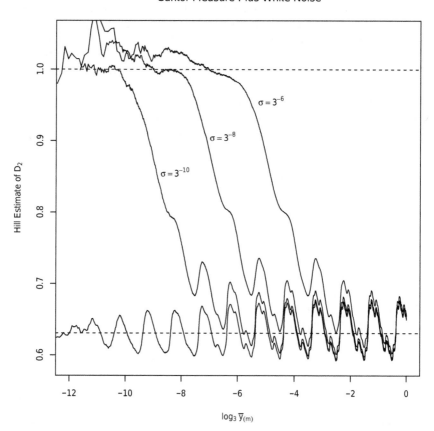

Figure 9.3 *Estimates of* D_2 *for the Cantor measure* ($p_0 = 0.5$) *with added white noise. The bottom line represents estimates for the pure Cantor measure. Lines above that represent estimates for the Cantor measure plus white noise with standard deviations of* 3^{-10}, 3^{-8} *and* 3^{-6}, *respectively. The horizontal dashed lines are at* $\log_3 2$ *and* 1, *the values of* D_2 *for the given Cantor measure and white noise, respectively* ($k = 100$, $n = 100{,}000$, $N = 100{,}000$).

9.4 Effect of Noise

We initially want to distinguish between two types of noise: observational error and system noise. Observational error occurs when a process x_i is recorded as $x_i + \epsilon_i$. System noise is more intrinsic to the process, and causes later observations (in time) to be functionally dependent on the noise inserted into the process at earlier steps, e.g., $x(t_i) = T_\xi[x(t_{i-1})] + \epsilon_i$. In this section we are interested only in observational noise.

The effect of adding observational error to data is one of the most serious forms of bias. Let the observational region be $\mathcal{X} \subseteq \mathbb{R}^d$, and the measure μ be concentrated on a subset of $\mathcal{X}$ of possibly lower dimension than d. Addition of error to observations tends to blur out any fine structure or to fill up the space, causing the dimension estimates to increase to d as the interpoint distances decrease to zero.

9.4.1 Cantor Measure

Consider the case where the Cantor measure was generated as in Example 8.6.2. Subsequently, white noise has been added. Results are plotted in Figure 9.3. As the standard deviation of the white noise increases, fine structure is progressively lost, hence for smaller interpoint distances, the noise dominates, and the dimension estimates are characteristic of the noise rather than the Cantor measure. It can also be seen that the effect of noise is much more serious than the rounding effect, for example.   □

The problem of noise in data and its effect on dimension estimates has also been discussed by Smith (1992b).

CHAPTER 10

Applications of Dimension Estimation

10.1 Introduction

In this chapter we describe some uses of dimension estimation, though apply the techniques to data simulated from mathematical and statistical models rather than real data. Using data with at least partially understood properties enables us to more easily evaluate the methods and interpret the results in real situations.

The next section contains two remaining estimation issues. The first is, given point estimates of $\theta(q)$, how does one estimate the multifractal spectrum $f(y)$? The second issue relates to lacunarity. In the Cantor measure examples given in Chapter 8, the lacunary cycle extended over the entire range of the interpoint distance y, and there was no boundary effect. It is wrong to assume that this is always the case for self-similar measures. An example will be given where there is both lacunarity together with a boundary effect. In the given example, they are both intrinsic forms of bias.

Examples of spatial point patterns that have features found in earthquake data are discussed in §10.3. One example is given where there is powerlaw scaling over much of the range of the interpoint distances, though the exponent is not the Rényi dimension. The other presents some features of the Moran cascade process.

In §10.4, two examples of dynamical systems are given and Rényi dimensions estimated. These are calculated under the assumption that the full phase space is observable. Often it is the case that the full phase space is not observable. In this situation, a reconstructed phase space is created that has the same fractal properties as the underlying phase space. These methods are briefly discussed.

Given an observed time series, a question which is of interest is whether the data were generated by a dynamical system of finite dimension or whether the system is stochastic. This begs the question of what is the difference between a deterministic and stochastic process, and is it possible to make this distinction based on empirical observations. These questions are discussed in §10.5.

Self-similar stochastic processes also satisfy various powerlaw relationships, in particular, the autocorrelation function, and the manner in which the distribution of the increment size scales with time. These processes have been discussed extensively in the literature since Lamperti's (1962) paper. However, these processes as originally defined were monofractal (one powerlaw index). More recently, these processes have been extended to 'multifractal' stochastic processes that satisfy a multiple scaling law as is the case for a multifractal measure. An introduction to both self-similar and multifractal stochastic processes is given in §10.6.

10.2 More on Estimation and Interpretation

In the first example of this section it is shown that dimension estimates calculated from observations sampled from a distribution (multinomial measure) on a self-similar set can in fact have both lacunary and boundary effects. In the previous chapters we studied the Cantor measure which only gave rise to a lacunary effect. Information about the way in which the measure was constructed can be deduced from the form of the boundary effect.

In the second example, an estimate of the multifractal spectrum is calculated using the Legendre transform based on estimates of the Rényi dimensions.

10.2.1 Example - Multinomial Measures

Measures that are supported on a self-similar set with all similarity ratios the same, and constructed by allocating uniform weight to each subset at each division (e.g., Moran cascade process where the similarity ratios are all equal) appear to give rise to a probability distribution, $F_Y(y)$, that is lacunary, where Y is the qth order interpoint distance. In the case of the Cantor measure, this lacunary behaviour is the only form of intrinsic bias (see Figure 7.4), and there is not an overall boundary effect as for the uniform distribution as shown in Figure 7.2, i.e., a gradual decrease in $\Phi(y)$ over the entire range of y. This lack of a boundary effect is caused by the placement of 'gaps'. Multinomial measures without such 'gaps', as defined in Theorem 5.5.2, not only have a lacunary effect but also an overall boundary effect.

In Figure 10.1, dimension estimates of simulated samples ($N = 100,000$) drawn from various multinomial measures are plotted. The measure parameters are written at the top of each graph where $p = (p_0, p_1, \cdots, p_{b-1})$. Each sampled point is generated by simulating a b-adic sequence, each digit with probability of p_ω of being ω, where $\omega \in \{0, 1, \cdots, b-1\}$. This is then converted to a decimal number between zero and one. One hundred bootstrap samples ($k = 100$) of $n = 100,000$ interpoint distances were simulated.

When the 'gap' occurs at the beginning or end, i.e., $p_0 = 0$ or $p_{b-1} = 0$, then it appears that the last lacunary cycle is incomplete, see the top left and bottom left plots. When there are no 'gaps', $F_Y(y)$ is lacunary but also has a pronounced boundary effect, see the top right plot. When there are 'gaps', there appears to be no boundary effect, see the bottom right plot. The boundary effect appears in varying strengths if only some 'gaps' are present, for example, compare the two plots in the middle row. Note that both of these cases have the same Rényi dimension $\tilde{D}_2$, and probably the same D_2, but the functions $F_Y(y)$ for each are clearly not the same. Theorem 5.5.4, which states that $D_2 = \tilde{D}_2$, is only applicable to a multinomial measure with gaps.

Note that the lacunary effect is manifested in $F_Y(y)$ as a periodic component with a period of one on a logarithmic scale of base b. In the construction of each of these multinomial measures, the similarity ratio was $1/b$. In general, self-similar

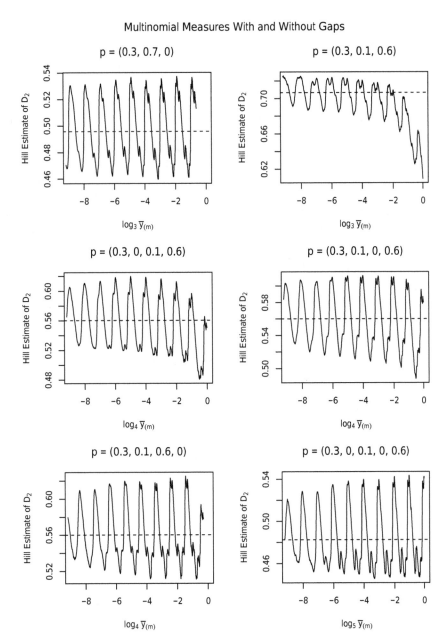

Figure 10.1 *Estimates of D_2 for various cases of the multinomial measure with and without 'gaps' where $p = (p_0, p_1, \cdots, p_{b-1})$. The horizontal dashed line is $\tilde{D}_2 = -\log_b \sum_{\omega=0}^{b-1} p_\omega^2$, ($k = 100$, $n = 100{,}000$, $N = 100{,}000$).*

sets can contain a number of different similarity ratios. When a multifractal measure is constructed on such a set, e.g., Moran cascade measure, there is a mixture of lacunary cycles that are out of phase with each other. As such, the collection of these lacunary cycles has the appearance of noise in dimension estimates of D_2, see Example 10.3.1.   □

10.2.2 Example - Cantor Measure

Consider again the case where $p_0 = 0.2$ with dimension estimates plotted in Figure 8.6 and corrected estimates of the Rényi dimensions given in Table 8.1. We will denote these as $\widehat{D}_q$, for $q = 2, \cdots, 8$, and the corresponding point estimates of $\theta(q)$ will be denoted as $\widehat{\theta}(q) = (q - 1)\widehat{D}_q$. Here we attempt to use these point estimates to partially reconstruct the multifractal spectrum $f(y)$ given by Equation 2.9. We do this by fitting a function to the point estimates $\widehat{\theta}(q)$, for $q = 2, \cdots, 8$. We will refer to this fitted function as $\widehat{\theta(q)}$. We then attempt to take the Legendre transform of $\widehat{\theta(q)}$ to give a functional estimate of $f(y)$, denoted as $\widehat{f(y)}$.

The Legendre transform of $\theta(q)$ is $f(y) = \inf_q \{qy - \theta(q)\}$. The solution can be shown to be $f(y(q)) = qy(q) - \theta(q)$, where $y(q)$ is the derivative of $\theta(q)$. Since

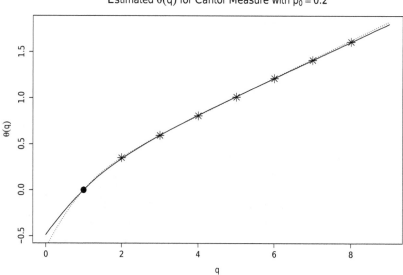

Estimated θ(q) for Cantor Measure with $p_0 = 0.2$

Figure 10.2 *The solid line is* $\widehat{\theta(q)}$ *for the Cantor measure with* $p_0 = 0.2$. *The stars represent the point estimates* $\widehat{\theta}(q)$, *for* $q = 2, \cdots, 8$, *and the solid dot represents the fixed point where* $\theta(1) = 0$. *The dotted line represents the function* $\widetilde{\theta}(q)$ *given by Equation 3.8.*

$\theta(q)$ is a rescaled cumulant generating function, its derivative will have the form given by the graph in Figure 3.3. That is, the derivative of $\theta(q)$ will decrease to a horizontal asymptote for increasing q, and also increase to a horizontal asymptote for decreasing q (see Figure 3.4). The form of the derivative function in Figure 3.3 is very similar to the logistic function. Assume that the lower asymptote is given by α_1 and the upper asymptote by α_2. Modifying the logistic function gives a possible parametric form for the functional estimate of $y(q)$ as

$$\widehat{y(q)} = \frac{\alpha_1 e^{-q}}{e^{-q} + 1} + \frac{\alpha_2 e^q}{e^q + 1}.$$

We then integrate to find the functional form of $\widehat{\theta(q)}$. Constants are added so that $\widehat{\theta(1)} = 0$ to give

$$\widehat{\theta(q)} = \alpha_1 \log(e^{-1} + 1) - \alpha_2 \log(e^1 + 1) - \alpha_1 \log(e^{-q} + 1) + \alpha_2 \log(e^q + 1).$$

If one had more point estimates of $\theta(q)$, additional parameters could be added to modify the rate at which the function approaches the two asymptotes. With the data we currently have, a two parameter model appears to be sufficient.

Non-linear least squares was used to estimate α_1 and α_2 giving estimates of $\widehat{\alpha}_1 = 0.964140$ and $\widehat{\alpha}_2 = 0.195159$. Using Equations 3.15 and 3.16, the true

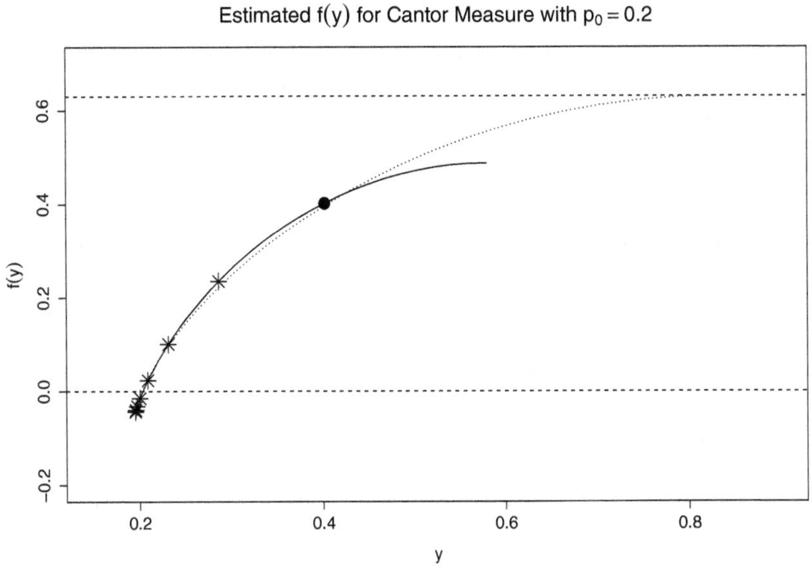

Estimated f(y) for Cantor Measure with $p_0 = 0.2$

Figure 10.3 *The solid line is $\widehat{f(y)}$ for Cantor measure with $p_0 = 0.2$. The solid dot represents the fixed point at $\widehat{y(1)}$, and the stars occur at $\widehat{y(2)}, \cdots, \widehat{y(8)}$ (top to bottom). The dotted line represents the true multifractal spectrum $f(y)$ given by Equation 3.6. The upper dashed line represents $f(y(0)) = D_0$.*

values of α_1 and α_2 can be calculated giving 1.464974 and 0.203114 respectively. The estimate of α_1 is poor because there are no data for $q \leq 0$.

The function $\widehat{\theta(q)}$ is plotted in Figure 10.2 for $0 < q < 9$, and is compared with the known values of $\widetilde{\theta}(q)$. The estimate of the multifractal spectrum is calculated as

$$\widehat{f(\widehat{y(q)})} = q\widehat{y(q)} - \widehat{\theta(q)},$$

which is plotted in Figure 10.3 and is compared with the known values of $\widetilde{f}(y)$. Note, as in Figure 3.2, $f(y)$ is positive. The negative values of $\widehat{f}(y)$ are caused by the point estimates $\widehat{\theta}(q)$ being too small for larger q, see Figure 10.2. □

10.3 Spatial and Temporal Point Patterns

Methods involving the estimation of exponents, like the correlation dimension, have often been used in point process analyses to describe the degree of clustering in spatial point patterns. For example, Kagan & Knopoff (1980) calculated what they called the two point correlation function to describe earthquake clusters. This was essentially the same as the correlation dimension D_2. Kagan (1981a, b) extended this to the three and four point correlation functions. These were not exactly the same as D_3 and D_4. The three point function was based on areas of triangles defined by the three points, and the four point function on volumes. The powerlaw number of these triangles or tetrahedrons was then determined as a function of their size δ. Kagan then normalised the estimates by the values of these functions when applied to the simple Poisson distribution.

Powerlaw estimates may not necessarily have a 'fractal' interpretation. The fractal dimension of a set describes the powerlaw increase in the required number of covers for the set as the cover width becomes infinitely small. When a set is self-similar, this powerlaw behaviour extends over all scales. In other situations, the powerlaw behaviour may only extend over an intermediate range of scales, and may breakdown on smaller and larger scales. This will be seen in Example 10.3.2, where D_2 is based on pairs of points sampled from the beta distribution, say X_1 and X_2. In this situation, D_2 is one, though parameters of the beta distribution can be selected so that $F_Y(y)$, where $Y = |X_1 - X_2|$, is powerlaw over almost the entire range of y with an exponent whose value is considerably smaller than one.

The cause of a powerlaw relationship not extending to the infinitely small scale may be due to data deficiencies, though powerlaw behaviour may be quite a sensible model assumption down to infinitely small scales. Alternatively, powerlaw behaviour down to a smaller and smaller scale may not be a plausible model assumption. When one calculates 'dimensions' one needs to be clear whether the interpretation of these numbers is in fact a dimension, implying a powerlaw exponent down to finer scales ad infinitum, or whether it is only powerlaw over some finite interval. In the later case, the exponent really relates to the probability distribution of the interpoint distances. In this section we further discuss some of these questions in relation to spatial point patterns and point processes.

10.3.1 Example - Moran Cascade Process

Spatial point patterns can be simulated with a probability distribution given by that of the Moran cascade process as follows. Consider the case in Example 6.2.3. The *seed* circle had radius one and was centred at $(0,0)$. Within this circle were placed four smaller circles with centres at $(V_0, W_0) = \left(\frac{1}{2}, 0\right)$, $(V_1, W_1) = \left(0, -\frac{2}{3}\right)$, $(V_2, W_2) = \left(-\frac{1}{2}, 0\right)$, and $(V_3, W_3) = \left(0, \frac{2}{3}\right)$, with similarity ratios $t_0, \cdots, t_3$ respectively. Here $b = 4$ and $\Omega = \{0, 1, 2, 3\}$. Given a random sequence $(\omega_1, \omega_2, \cdots) \in \Omega^\infty$, where $\omega_i \in \Omega$, and using the coding map in Equation 6.1, the coordinates of the simulated point can be determined. Digits $j \in \Omega$ are simulated with probability p_j such that $\sum_{j \in \Omega} p_j = 1$. For our simulations, a sequence of length $n = 40$ is sufficient, and then coordinates of the simulated point are given by

$$x = V_{\omega_1} + \sum_{k=1}^{n-1} \left(V_{\omega_{k+1}} \prod_{i=1}^{k} t_{\omega_i} \right)$$

and

$$y = W_{\omega_1} + \sum_{k=1}^{n-1} \left(W_{\omega_{k+1}} \prod_{i=1}^{k} t_{\omega_i} \right).$$

Simulated Moran Cascade Processes

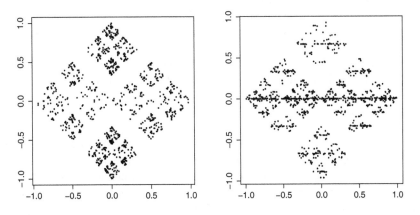

Figure 10.4 *Simulated Moran cascade processes* ($N = 1000$) *with similarity ratios* $t_0 = \frac{1}{2}$ *(right),* $t_1 = \frac{1}{3}$ *(bottom),* $t_2 = \frac{1}{2}$ *(left), and* $t_3 = \frac{1}{3}$ *(top). The left-hand graph has all* $p_j = 0.25$, *whereas the right has* $p_0 = p_2 = 0.4$ *and* $p_1 = p_3 = 0.1$.

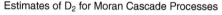

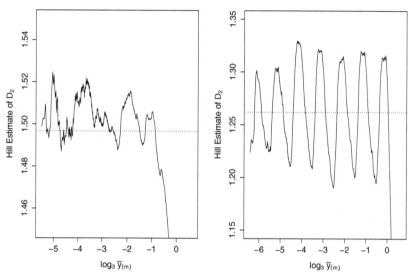

Figure 10.5 *Estimates of D_2 for a simulated Moran cascade processes of length $N = 10{,}000$. In both cases $p_j = 0.25$, $j = 0, \cdots, 3$. The horizontal dotted lines are $D_2^{\star}$. The left-hand graph is the same process as in the left-hand graph of Figure 10.4 where the similarity ratios are $t_0 = \frac{1}{2}$ (right), $t_1 = \frac{1}{3}$ (bottom), $t_2 = \frac{1}{2}$ (left), and $t_3 = \frac{1}{3}$ (top). The right-hand process has all similarity ratios constant, i.e., $t_j = \frac{1}{3}$ for $j = 0, \cdots, 3$.*

Figure 10.4 shows two simulations, each of 1000 points, with different values of the p_j's. The first graph has all $p_j = 0.25$ thus giving each 'cluster' the same expected number of points. The higher density of the upper and lower clusters is caused by the smaller similarity ratios. The second process gives greater probability to the left and right clusters, giving the appearance of many points aligned in the middle with smaller parallel alignments. The points may represent initiation points of micro stress fractures in a two-dimensional object. The alignment of many such micro fractures has produced a main 'fault' with smaller parallel 'faults'.

The graph on the left-hand side of Figure 10.5 shows the dimension estimates of D_2 for the same process plotted on the left of Figure 10.4 but with a sample size of $N = 10{,}000$. Note that this process has a mixture of similarity ratios and hence there is not a clear lacunary cycle. The dimension estimates on the right-hand side of Figure 10.5 represent the same process, except all similarity ratios are $\frac{1}{3}$. This produces a clear cycle which is periodic on a logarithmic scale of base 3. Note how the boundary effect occurs later in the case where the similarity ratios are the same (lacunary cycles effectively in phase with each other).   □

Dimension estimates of spatial patterns can sometimes be dominated by two clusters of points that are much larger than other clusters. This is seen in the following example where apparent powerlaw behaviour is produced in the estimates of D_2, though one gets humps in the estimates of the higher order Rényi dimensions.

10.3.2 Beta Distribution

The beta distribution can be used to provide an interesting antithesis to the boundary effect discussed in §9.2. The boundary effect is caused by a relative paucity of larger interpoint distances. If we sample pairs of points from a beta distribution with $\alpha = \beta < 1$, we will tend to sample more larger interpoint distances than when $\alpha = \beta = 1$ (i.e., uniform) because the mass is more concentrated towards the edge of the unit interval as α and β decrease.

Further, the probability distribution of $Y = |X_1 - X_2|$ becomes increasingly powerlaw like with exponent 2α over much of the range of Y. This can be seen in the plots of Figure 10.6. When $\alpha = \beta = 0.5$, the boundary effect in the estimates of D_2 are much weaker than in the case where $\alpha = \beta = 1$ (i.e., uniform). In the cases for still smaller values of $\alpha = \beta$, the line representing estimates of D_2 indicates that $F_Y(y)$, with $q = 2$, has powerlaw behaviour over almost its entire range.

This powerlaw behaviour can be verified using a known result on the difference of two beta random variables. In our case $\alpha = \beta$ so the distribution of the difference will be symmetrical about zero. It follows from Johnson, Kotz & Balakrishnan (1994, Equation 25.103e) that the probability density function of Y, $f_Y(y)$, has the form

$$f_Y(y) = \frac{y^{2\alpha-1}(1-y)^{2\alpha-1}}{B(\alpha,\alpha)} \sum_{i=0}^{\infty} \sum_{j=0}^{\infty} A_{\alpha,i,j} \frac{(1-y)^i(1-y^2)^j}{i!\,j!},$$

where $B(\alpha,\beta)$ is the beta function and $A_{\alpha,i,j}$ are non-negative constants. The infinite series is convergent for $y > 0$. By expressing $(1-y)^{2\alpha-1}$ as a Taylor's series ($y < 1$), the density function can be expressed as a polynomial with leading term $y^{2\alpha-1}$. Integrating shows that the probability distribution of Y is powerlaw with exponent 2α for sufficiently small $\alpha = \beta$. The smaller the value of $\alpha = \beta$, the more dominant this leading term becomes. When α and β are sufficiently small, the beta distribution is essentially two powerlaw distributions at opposite ends of the unit interval.

The fact that the beta density is not multifractal is indicated by the higher order Rényi dimension estimates. None of these are powerlaw in the manner of D_2. Further, as q increases there is an increasing single peak, indicating that the interplay is essentially between two clusters.   □

The above example has been included because it is not completely dissimilar to the behaviour that can occur with earthquake spatial point patterns, where two

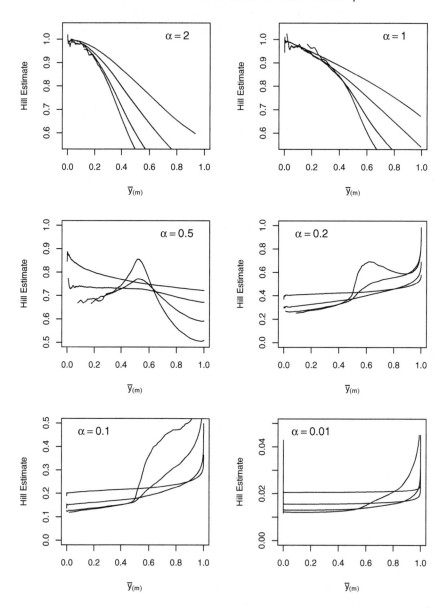

Figure 10.6 *Estimates of D_2, D_3, D_5 and D_7 when points are sampled from a beta distribution with $\alpha = \beta$. In the top two plots, lines represent D_2, D_3, D_5 and D_7 top to bottom. In the bottom four plots, lines for small values of y represent D_2, D_3, D_5 and D_7 top to bottom, though ultimately the most peaked is that of D_7.*

very densely populated clusters can dominate the dimension estimates. This will be discussed further in Chapter 11.

10.3.3 Consistency for Point Processes

When dimension estimates are calculated from an observed point pattern, there is some ambiguity as to the interpretation of the estimates. A finite point pattern itself has dimension zero. In the case of a point pattern simulated from the Moran cascade process, for example, the Rényi dimension estimates relate to the probability distribution from which the data were sampled.

Vere-Jones (1999) shows that there are at least two different interpretations of the Rényi dimension estimates in the case of an observed space-time point process. In the first, the spatial region is fixed and estimates are taken over increasing periods of time, and in the second, the time interval is fixed while the size of the spatial region increases. These two situations are discussed further in §11.1.1.

10.4 Dynamical Systems

In this section we estimate the Rényi dimensions of the logistic map and the Lorenz attractor. The Cantor like structure in the example of the logistic map is clearly evidenced by the lacunary behaviour of the dimension estimates.

Generally only a projection of a dynamical system is observed, often only a scalar time series. That is, the full process is not completely observable. Using the observed time series, one attempts to determine the multifractal characteristics of the full unobserved process. This is discussed and applied to the Lorenz attractor.

10.4.1 Logistic Map

We initially discussed the logistic map in §1.3.2. The behaviour of the logistic map undergoes a sequence of bifurcations as ξ increases to $\xi_\infty \approx 3.569945672$, see §1.3.2. When $\xi = \xi_\infty$ (see Figure 1.4), the attracting set is similar in nature to the Cantor set.

In this example, we estimate the Rényi dimensions ($q = 2, \cdots, 5$) for the Cantor like case where $\xi = \xi_\infty$. We use $\xi = 3.569945672$ as an approximation for ξ_∞ in our simulations. Using an approximation to ξ_∞ causes interesting behaviour. If we simulate a sequence of sufficient length, it will start repeating its orbit, even though the period is very large (see Grebogi et al., 1988). This will cause the dimension estimates to tend to zero for small interpoint distances. When the simulated series is sufficiently short that no periodic behaviour is detected, the approximation to ξ_∞ is more similar to adding noise to the simulated series, and the dimension estimates will tend to one for small interpoint distances.

If the simulated series starts repeating itself, there is a reasonable likelihood that zero interpoint distances will be sampled. The Hill estimate given by Equation 8.8 cannot handle zero distances. In practice, we use the modified Hill estimate given in §9.3.2 and assume that the machine precision is approximately 10^{-16}. This value is then used for h in §9.3.2.

The above effects can be seen in Figure 10.7. Five values of ξ, close to ξ_∞, have been used to simulate sequences of length $N = 100,000$. For each sequence, the Hill estimates of D_2 have been calculated using $k = 100$ bootstrap samples each consisting of $n = 100,000$ sampled interpoint distances. The interpoint distances on the horizontal axis have been plotted on a logarithmic scale of base 6, because this is the approximate self-similar scaling factor determined from Figure 1.3. In the cases where $\xi = 3.568$ and 3.569, zero interpoint distances were sampled, and hence the dimension estimates for small values of the interpoint distance are zero. For the other values of ξ, the process exhibits Cantor like behaviour for larger interpoint distances. This is seen by the development of lacunary like behaviour close to the Hausdorff dimension (dotted line) of the attractor (Falconer, 1990, page 176). The dimension estimates tend to one for smaller interpoint distances, i.e., the fine structure becomes blurred due to the approximation used for ξ_∞.

Dimension estimates of $D_2, \cdots, D_5$ for the case where $\xi = 3.569945672 \approx \xi_\infty$ are plotted in Figure 10.8. In the case of the logistic map, an extremely long series needs to be simulated before lacunary like behaviour becomes evident over a few cycles. For the calculations in Figure 10.8, a series of length $N = 300,000$ was simulated, and $k = 100$ bootstrap samples selected with each

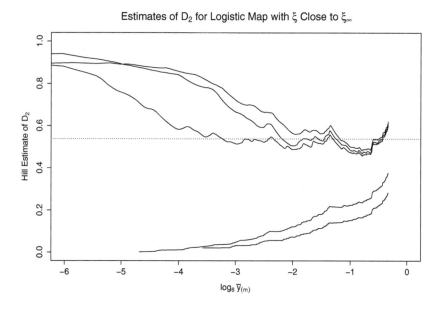

Figure 10.7 *Estimates of D_2 for the logistic map with $\xi = 3.568$ (bottom line), 3.569 (2nd to bottom), 3.570 (3rd down), 3.571 (2nd down), 3.572 (top)* ($k = 100$, $n = 100,000$, $N = 100,000$).

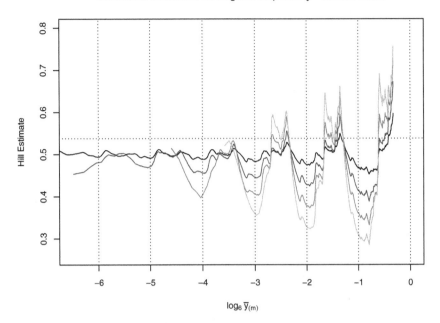

Figure 10.8 *Estimates of $D_2, \cdots, D_5$ for the logistic map with $\xi = 3.569945672$. The dotted line represents the Hausdorff dimension $(0.538 \cdots)$ of the attractor (Falconer, 1990, page 176). The line with the smallest peak and most shallow trough represents the estimate of D_2, etc., to the line with the greatest peak and deepest trough represents D_5 $(k = 100, n = 300,000, N = 300,000)$.*

sample consisting of $n = 300,000$ interpoint distances. Note that the lacunary cycle appears to be approximately periodic on a logarithmic scale with a base of six, consistent with Figure 1.3. The resolution is relatively poor compared to the estimates of the Cantor measure in Figures 8.5 and 8.6, because the attracting set is contained in a much smaller part of the unit interval (see Figure 1.4), and an approximate value of ξ_∞ was used in the simulations. The line representing the dimension estimate of D_2 extends back almost horizontally at about 0.5 until about -7.0, then it tends up to 1. This is again consistent with the error induced by using an approximation for ξ_∞.

Using Figure 10.8, a rough estimate of D_2 is 0.5. Estimates of D_3, D_4 and D_5 only decrease small amounts compared to the Cantor Measure where $p_0 = 0.2$ shown in Figure 8.6, with D_5 for the logistic map being not much less than 0.48. This would tend to indicate that the orbit of the logistic map visits various subsets of the attracting set more evenly than in the case of the Cantor map with $p_0 = 0.2$ (see Example 1.3.1). $\square$

10.4.2 Generalised Bakers' Map

The bakers' map is so named because it consists of a stretching and then a folding transformation, similar to the operations performed by a baker preparing dough. This is discussed by Falconer (1990, page 177). Smith (1992a) gives a generalisation of this map called the *generalised bakers' map* and calculates the corresponding function $\Phi(y)$ when $q = 2$. In this case, $\Phi(y)$ has very similar properties to that for the Cantor measure in Figure 7.5.   □

10.4.3 Lorenz Attractor

The Lorenz equations are defined in $\mathbb{R}^3$ as in Example 1.3.3. This is referred to as the *phase space*, within which the complete system of equations are defined. When fractal dimensions are being estimated for an observed dynamical system, it is often assumed that only part of the process is observable, that is, we observe the process in an *observation space*. Hence there is an assumption of a projection from the phase space to the observation space, where our observations may even consist of only a scalar time series.

In this example we assume that the observation and phase spaces are the same, and that our observed point locations of the orbit include all components of the

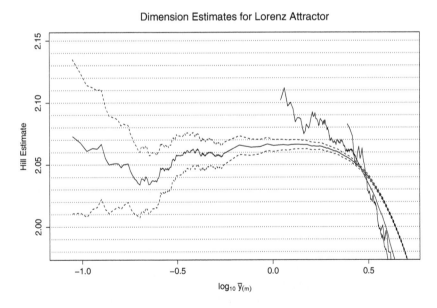

Figure 10.9 *Estimates of $D_2, \cdots, D_5$ (solid lines) for the Lorenz attractor. The line representing D_2 extends furthest to the left, then followed by D_3, D_4 and D_5. The dashed lines represent an interval centred on the estimates of D_2 with a half width of twice the standard error ($k = 100$, $n = 200{,}000$, $N = 200{,}000$).*

three dimensions. We will consider in §10.4.4 what happens when the phase and observation spaces are not the same.

In this example, we have used the approximation given by Equation 1.1 (Euler's method) to simulate a series using a step length of $h = 0.001$. The number of simulated observations (in $\mathbb{R}^3$) was $N = 2 \times 10^6$. Dimension estimates were calculated by sampling $n = 2 \times 10^5$ interpoint distances and repeating for $k = 100$ bootstraps. Hill estimates are plotted in Figure 10.9.

Intervals representing twice the standard errors have also been plotted in Figure 10.9 about the Hill estimates of D_2. It is clearly seen that the variability of the estimates are greater for smaller interpoint distances but also the autocorrelation as m increases is also evident; compare with Figure 8.1.

Also note that the estimates of D_3 to D_5 have not extended to sufficiently small interpoint distances where powerlaw behaviour is evident. This is a problem that occurs in estimating higher order Rényi dimensions. Long series of observations are required before dimension estimates are not affected by the boundary and fall within the range where powerlaw scaling is evident. □

10.4.4 Embedding and Reconstruction

In our discussion up to this point on dynamical systems, we have assumed that the process is directly observable on $\mathcal{X} \subseteq \mathbb{R}^d$, and can be represented as a discrete time mapping

$$x(t_{n+1}) = T_\xi[x(t_n)] = T_\xi^{n+1}[x(t_0)].$$

Physicists refer to $\mathcal{X}$ as the *phase space*. The purpose has been to describe the probability measure μ, which is the probability of finding the trajectory within the set A at any given time point. Since the process is deterministic and the measure μ is assumed to be invariant under the mapping T_ξ, $\mu(A)$ is equivalent to the probability that $x(t_0) \in A$. The condition that μ is invariant is equivalent to the time series $\{x(t_n) : n = 0, 1, \cdots\}$ being strictly stationary.

Often the phase space, $\mathcal{X}$, of a dynamical system is not directly observable, and we only observe the process in another space, say $\mathcal{Y} \subseteq \mathbb{R}^k$ where $k < d$, which is a projection of $\mathcal{X}$, i.e.,

$$\Pi : \mathcal{X} \longrightarrow \mathcal{Y}.$$

We will refer to $\mathcal{Y}$ as the *observation space*. Often $k = 1$, and then our observations consist of a univariate time series $\{y(t_n) : n = 0, 1, \cdots\}$. Given these observations, can we still determine the multifractal characteristics of the probability measure μ?

The multifractal characteristics can often be determined by embedding the time series into a p-dimensional *reconstructed phase space*. This is usually done by the *method of time delays*, where a new p-dimensional time series, denoted by

$y^{(p)}(t_n)$, is defined as

$$y^{(p)}(t_n) = \big(y(t_n), y(t_{n+\ell}), \cdots, y(t_{n+(p-1)\ell})\big), \qquad (10.1)$$

where $\ell \in \mathbb{Z}^+$ is a lag parameter. The required value of p will be discussed below. Now by re-expressing $y^{(p)}(t_n)$ as

$$y^{(p)}(t_n) = \Big(\Pi \circ T_\xi^n[x(t_0)], \Pi \circ T_\xi^{n+\ell}[x(t_0)], \cdots, \Pi \circ T_\xi^{n+(p-1)\ell}[x(t_0)]\Big),$$

it is seen that we have a composite mapping, say Ψ_p, as

$$\Psi_p : \mathcal{X} \longrightarrow \mathcal{Y}^p, \qquad (10.2)$$

where $\mathcal{Y}^p$ is the product space of $\mathcal{Y}$. We say that Ψ_p is an *embedding map* if the fractal properties of $\mathcal{X}$ are preserved in the reconstructed phase space $\mathcal{Y}^p$. Usually the notion of embedding includes the requirement that differential structure is also preserved, though this is not necessary here.

The method of time delays, as in Equation 10.1, is based on a result by Takens (1981), which is based on the Whitney (1936) embedding theorem. Such time delay reconstructions were first advocated by Packard et al. (1980). Sauer et al. (1991) showed that if p in Equation 10.1 is greater than twice the box counting dimension of the attractor in $\mathcal{X}$, then Ψ_p will be an embedding map.

Further discussion on the notion of embedding can be found in Whitney (1936), Sauer et al. (1991), Cutler (1997), Cheng & Tong (1994), and Isham (1993). Casdagli et al. (1991) discuss state space reconstruction in the presence of noise, and Fraser & Swinney (1986) discuss optimal time lag length for reconstruction. Further discussion can also be found in Cutler & Kaplan (1997) which contains a collection of review articles by both mathematicians and physicists. A review article on the analysis of observed chaotic data by Abarbanel et al. (1993), and the book by Abarbanel (1995), also contain excellent sections on reconstructing the phase space.

10.4.5 Example - Lorenz Attractor Continued

In this example, we assume that only the $x_1(t_n)$ component of the Lorenz attractor is observable, that is $y(t_n) = x_1(t_n)$ where $x_1(t_n)$ is defined in Equation 1.2. Using this scalar time series, we will form an embedding using the method of time delays, $y^{(p)}(t_n)$, into a reconstructed phase space whose fractal properties should be the same as those of $\mathcal{X}$ analysed in Example 10.4.3.

We know that p must be greater than twice the box counting dimension of the attractor. However, we also need to determine the lag parameter ℓ. The best value to select is not easy to prescribe. We are observing a system whose autocorrelations are of interest. However, if ℓ is too small, successive observations will add little information to our knowledge. For example, a time series of air temperature measurements in Wellington at one second intervals probably contains much redundant information. Alternatively, if ℓ is too large, we will loose the relationship between successive values, and the readings will be essentially independent.

Lorenz Attractor: Various Lag Lengths

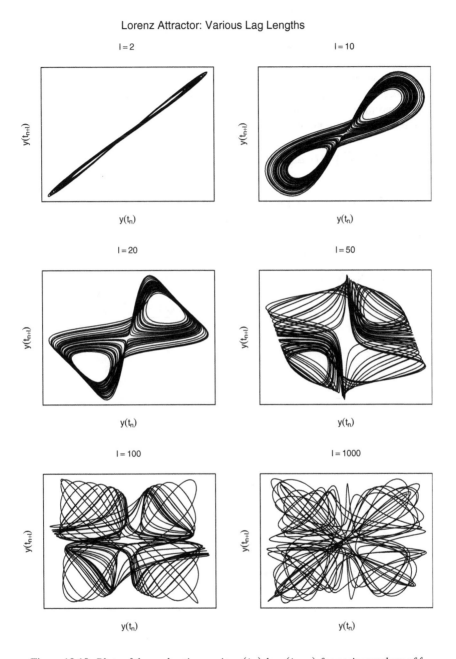

Figure 10.10 *Plots of the scalar time series* $y(t_n)$ *by* $y(t_{n+\ell})$ *for various values of* ℓ.

In Figure 10.10, the scalar time series $y(t_n)$ is plotted against $y(t_{n+\ell})$ for various values of ℓ. It appears that the structure of the attractor 'unfolds' when ℓ is somewhere between 10 and 20, compare with Figure 1.5. When ℓ is too large, the characteristic structure appears to be lost.

Fraser & Swinney (1986) suggested that ℓ could be determined by calculating the *average mutual information* as a function of ℓ, defined as follows. Let U and W be random variables taking values $u_1, u_2, \cdots$ and $w_1, w_2, \cdots$ respectively. Then the average mutual information is defined as

$$H = \sum_i \sum_j \Pr\{U = u_i, W = w_j\} \log_2 \frac{\Pr\{U = u_i, W = w_j\}}{\Pr\{U = u_i\} \Pr\{W = w_j\}},$$

where $\Pr\{U = u_i, W = w_j\}$ is the joint probability of U and W. In the case of a continuous scalar time series $Y(t_n)$, we can divide the range of $Y(t_n)$ into intervals, say $I_1, I_2, \cdots$. We are interested in the average mutual information between $Y(t_n)$ and $Y(t_{n+\ell})$ for various values of ℓ. Hence $H(\ell)$ is defined as

$$\begin{aligned} H(\ell) \;=\; & \sum_i \sum_j \Pr\{Y(t_n) \in I_i, Y(t_{n+\ell}) \in I_j\} \\ & \times \log_2 \frac{\Pr\{Y(t_n) \in I_i, Y(t_{n+\ell}) \in I_j\}}{\Pr\{Y(t_n) \in I_i\} \Pr\{Y(t_{n+\ell}) = w_j\}}. \end{aligned}$$

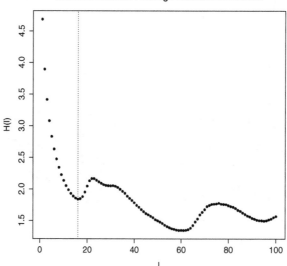

Lorenz Attractor: Average Mutual Information

Figure 10.11 *Average mutual information for the scalar series* $y(t_n) = x_1(t_n)$ *of the Lorenz attractor. The first minimum occurs at* $\ell = 16$.

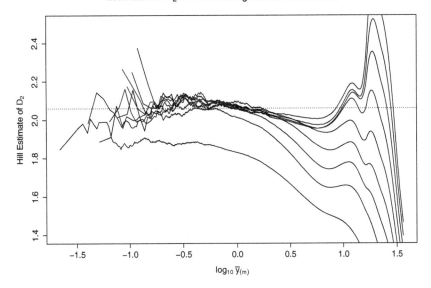

Figure 10.12 *Estimates of D_2 using the method of time delays for the Lorenz attractor, assuming that only the $x_1(t_n)$ component (as in Equation 1.2) is observed. The lines represent the dimension estimates using $y^{(p)}(t_n)$ for $p = 2, 3, \cdots, 10$. The horizontal dotted line is at 2.06.*

In the case of an observed time series $y(t_n)$, we define intervals I_i and estimate the probabilities as the proportion of times that the interval I_i is visited. Fraser & Swinney (1986) suggested that $H(\ell)$ be calculated for $\ell = 1, 2, 3, \cdots$, and that value of ℓ where the first minimum is found is to be used as the time lag value in the calculation of $y^{(p)}(t_n)$, given in Equation 10.1. This has been done in Figure 10.11, where a series of 20,000 was simulated and the range of $y(t_n)$ was divided into 200 subintervals. The first minimum occurs at $\ell = 16$.

Using a value of $\ell = 16$, $y^{(p)}(t_n)$ as in Equation 10.1 was calculated. Estimates of D_2 for $y^{(p)}(t_n)$ are plotted in Figure 10.12. The bottom line represents the dimension estimates of $y^{(2)}(t_n)$, the next of $y^{(3)}(t_n)$, etc., and the top line of $y^{(10)}(t_n)$. As p increases, the dimension estimates appear to saturate at about the same value of D_2 in Figure 10.9. One needs to be careful in concluding that the dimension estimates have saturated for a given value of p. As p increases, one needs an increasingly large sample size to get dimension estimates within the range where powerlaw behaviour is evident. See for example Figure 10.13 which plots estimates of D_2 for white noise in up to five dimensions. In the case of five dimensions, the line representing the dimension estimates only just reaches five.

□

10.5 Is a Process Stochastic or Deterministic?

Simple non-linear sets of equations can generate extremely complex behaviour, and often the dynamics appear stochastic. In many observed time series, it is not clear what the fundamental underlying process is that drives the system. For example, the climate in Wellington could be partially described by a time series of daily maximum temperatures, daily rainfall, or possibly even the velocity of the maximum daily wind gust. Alternatively, the movement of the earth's tectonic plates is made evident by recorded seismic events and surface deformation. However, in both processes, we only observe certain aspects that are evidence of an underlying more complex process, hence the notion of a projection as described in §10.4.4. A question that is of interest is, given a time series that is related to an underlying more fundamental process, is it possible to determine whether the underlying process is governed by a deterministic set of equations or a stochastic system.

Dimension estimation has been used as a method to determine whether a time series is stochastic or deterministic. Given a time series, typically univariate, a time delay embedding is calculated for $p = 2, 3, \cdots$ as in Equation 10.1. Usually the correlation dimension, D_2, is estimated for each value of p. If the time series is deterministic, then the estimates of D_2 should converge to a limit. If the time series is stochastic, then $D_2 = p$, and so the dimension estimates should increase indefinitely as the dimension of the reconstructed phase space increases. This is done under the assumption that a stochastic time series has an infinite number of degrees of freedom whereas that generated by a dynamical system has a finite number of degrees of freedom.

In an important paper by Osborne & Provenzale (1989), it was shown that a certain coloured noise process with powerlaw spectra saturated at a finite dimension, contrary to what was expected (see also Theiler, 1991 and Kennel & Isabelle, 1992). It was later shown by Cutler (1994) that their example failed to have an infinite dimension because the powerlaw spectrum was discrete. Subsequently, Cutler (1997) defined more precisely what is meant by stochastic and deterministic. We give a brief summary in this section.

10.5.1 Definitions (Cutler, 1997)

1. The time series $\{Y(t_n) : n = 0, 1, \cdots\}$ is said to be *strictly stationary* if for any finite collection $t_1, \cdots, t_n$ and for all τ,

$$\Pr\{Y(t_1) < y_1, \cdots, Y(t_n) < y_n\}$$
$$= \Pr\{Y(t_1 + \tau) < y_1, \cdots, Y(t_n + \tau) < y_n\}.$$

2. A mapping $g : \mathcal{X} \rightarrow \mathcal{Y}$, between the metric spaces $\mathcal{X}$ and $\mathcal{Y}$ with metrics ρ_1 and ρ_2 respectively, is said to satisfy a *Lipschitz condition* if, for all $x_1, x_2 \in \mathcal{X}$,

$$\rho_2(g(x_1), g(x_2)) \leq k\rho_1(x_1, x_2),$$

where k is a constant. If in addition, g is one to one and g^{-1} also satisfies a Lipschitz condition on its domain, then g is bi-Lipschitz.

3. Let $\{Y(t_n) : n = 0, 1, \cdots\}$ be a strictly stationary time series with values in $\mathcal{Y}$. The *predictive dimension*, denoted by ζ, is defined as the smallest $n \geq 1$ such that there exists a mapping $\Upsilon : \mathcal{Y}^n \longrightarrow \mathcal{Y}$ such that

$$Y(t_n) = \Upsilon [Y(t_0), \cdots, Y(t_{n-1})]$$

with probability 1. If no function Υ exists for all $n \geq 1$, then $\zeta = \infty$.

□

The following theorem tells us that a strictly stationary process with known predictor function Υ and finite predictive dimension ζ can be predicted as a function of the previous ζ observations. This is then used to define what we mean by a stochastic and deterministic time series.

10.5.2 Theorem (Cutler, 1997, Theorem 2.2)

Let $\{Y(t_n) : n = 0, 1, \cdots\}$ be a strictly stationary time series with finite predictive dimension ζ and predictor function Υ. Then, for all integers $m \geq 0$,

$$Y(t_{m+1+\zeta}) = \Upsilon (Y(t_{m+1}), Y(t_{m+2}), \cdots, Y(t_{m+\zeta}))$$

with probability 1.

□

10.5.3 Definition (Cutler, 1997)

A strictly stationary time series $\{Y(t_n) : n = 0, 1, \cdots\}$ is said to be *deterministic* if $\zeta < \infty$ and *stochastic* if $\zeta = \infty$, where ζ is the predictive dimension. □

The following theorem tells us that, under certain conditions, the Rényi dimensions associated with a dynamical system with invariant measure μ are the same as those associated with the time lagged process $Y^{(p)}(t_n)$ with distribution P_n for suitable values of p.

10.5.4 Theorem (Cutler, 1997, Corollary 2.26)

Let $T : \mathcal{X} \to \mathcal{X}$ be a dynamical system with invariant distribution μ, i.e., $X(t_n) = T^n(X(t_0))$, for $n = 1, 2, \cdots$, where $X(t_0)$ is a random initial condition with distribution μ. Suppose that the projection $\Pi : \mathcal{X} \to \mathcal{Y}$ is Borel measurable and $Y(t_n) = \Pi[T^n[X(t_0)]]$ is the resulting functional time series with joint distributions P_n, i.e.,

$$P_n(y_0, \cdots, y_{n-1}) = \Pr\{Y(t_0) < y_0, \cdots, Y(t_{n-1}) < y_{n-1}\}.$$

Also assume that the delay coordinate mapping Ψ_p, given by Equation 10.2, is bi-Lipschitz for some p, and T satisfies a Lipschitz condition (Definition 10.5.1). Then

1. the functional time series $Y(t_n)$ is deterministic with predictive dimension $\zeta \leq p$;

2. for all $n \geq p$, $D_q(\mu) = D_q(P_n)$, for $q \in \mathbb{R}$, where $D_q(\nu)$ is defined as in Equation 2.6, but with μ replaced by the distribution ν; and

3. for all $n \geq p$, $\dim_H(\text{supp}\,\mu) = \dim_H(\text{supp}\,P_n)$.

□

10.5.5 Example - Gaussian Time Series

A scalar time Gaussian series $\{y(t_n) : n = 0, \cdots, 100{,}000\}$ was simulated, and the p-dimensional time series $y^{(p)}(t_n)$ was calculated using Equation 10.1 (note that $\ell = 1$ here). Figure 10.13 shows estimates of D_2 using the series $y^{(p)}(t_n)$ for $p = 1, \cdots, 5$. Since the process is stochastic, $D_2 = p$. However, note that as p increases, the dimension estimates are increasingly affected by the boundary effect, and for $p = 5$, there is no real powerlaw behaviour, with the estimates only just reaching five.

Wolff (1990) derived an expression for the correlation integral (see §7.1.2) with $q = 2$ for an autoregressive time series and a moving average time series. He also simulated AR(1) and MA(1) processes with Gaussian white noise innova-

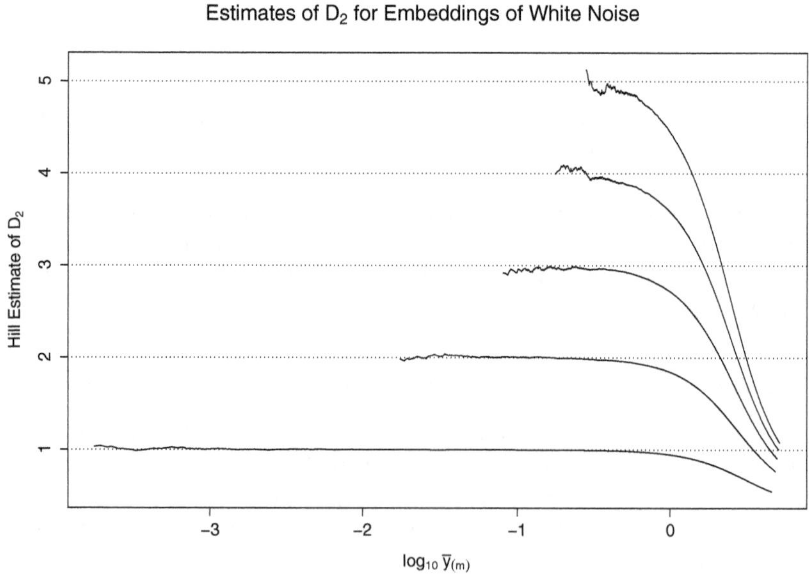

Estimates of D₂ for Embeddings of White Noise

Figure 10.13 *Estimates of D_2 for a scalar time series $y(t_n)$ of white noise, $n = 0, \cdots, 100{,}000$. The bottom line represents estimates using the series $y^{(1)}(t_n)$, the second of $y^{(2)}(t_n)$, and so on, where $y^{(p)}(t_n)$ is given by Equation 10.1. There were 100 bootstrap samples each consisting of 100,000 sampled interpoint distances.*

tions, and calculated the time delay embeddings $y^{(p)}(t_n)$. Since the processes are stochastic, D_2 should be equal to p, the dimension of the reconstructed phase space. In the simulations done by Wolff (1990), his estimates of D_2 were $\leq p$, with the difference becoming greater as p increased; i.e., as the boundary effect becomes more of a problem. The estimates of D_2 were also smaller when the simulated series had greater autocorrelations between successive observations.

These examples show that, while a stochastic process may have infinite degrees of freedom and hence is infinite dimensional, ones ability to detect this from an observed time series is hindered by the boundary effect. As p increases, this effect will become increasingly severe and can give the appearance that the dimension of the process is converging to a finite value as would be expected for a deterministic time series. ☐

10.6 Stochastic Processes with Powerlaw Properties

10.6.1 Historical Sketch

There are three important papers that provide the seminal ideas of processes that display powerlaw scaling and long range dependence (Hurst, 1951; Rosenblatt, 1961 and Lamperti, 1962). We briefly introduce the ideas contained in each.

Harold Edwin Hurst (1951) was a hydrologist studying discharge rates of the River Nile at Aswan. Given the inflow and outflows of the reservoir, he derived an expression for the storage range required by an ideal reservoir in the interval of time $(0, T)$, say $R(T)$, to ensure that it could cope with times of high input (i.e., does not overflow) and times of high demand. Thus $R(T)$ is the required capacity of the reservoir for it to function satisfactorily over an interval of length T. It is dependent on the interval length because the storage required can be thought of as being similar to the path of Brownian motion. The longer the period, the more variable the process becomes, and hence the greater the possible extremes. Hurst (1951) analysed the historical records and found that $R(T)/S$, where S is the standard deviation of $R(1)$, calculated from different parts of the historical record, was approximately equal to $(T/2)^H$. Hurst (1951) pointed out that if inflows and outflows were independent, similar to sums of mutually independent and identically distributed random variables, then one would expect R/S to follow a law of this type, but with $H = \frac{1}{2}$. He found that H was typically about 0.73. Hurst attributed this result to be a consequence of the flow rates having serial correlation. The *Hurst parameter* H now often appears in the formulation of stochastic processes that display long range dependence. Further information on the stochastic modelling of river flows can be found in Lawrance & Kottegoda (1977) and Lloyd (1981). So called R/S analyses have often been used to estimate the Hurst parameter, see for example Mandelbrot & Wallis (1969) for a description of the method, and Mandelbrot (1975) for various limit distribution results.

Rosenblatt (1961) was concerned with the distribution of sums of sequences of random variables that were strongly mixing. A sequence Y_n is said to be strongly mixing if

$$\big| \Pr\{Y_n \in A, Y_{n+m} \in B\} - \Pr\{Y_n \in A\} \Pr\{Y_{n+m} \in B\} \big| \leq d(m),$$

where $d(m)$ decreases to 0 as $m \to \infty$. He showed that if the sums of such sequences satisfied certain moment conditions, then their limiting distribution, when suitably normalised, should have a Gaussian distribution. He gave a counter example where the sequence of observations Y_n satisfied the moment conditions but were serially correlated. The normalised sums converged to to a non-Gaussian distribution. He concluded that the sequence Y_n did not satisfy the strong mixing condition. In fact it displayed a form of long range dependence.

The stationary sequence Y_n, with zero mean, is said to be in the *domain of attraction* of a process $X(t)$ if the finite dimensional distribution of

$$X_N(t) = \frac{1}{d_N} \sum_{j=1}^{\lfloor Nt \rfloor} Y_j \tag{10.3}$$

converges to those of $X(t)$ as $N \to \infty$, where d_N is any positive normalising factor which tends to infinity as $N \to \infty$, and $\lfloor Nt \rfloor$ denotes the largest integer not greater than Nt. Rosenblatt's (1961) counter example converged to a process that has become known as the *Rosenblatt process*.

The following discussion uses the notion of *equality of finite dimensional distributions*. We will say that two stochastic processes $X_1(t)$ and $X_2(t)$ have the same finite dimensional distributions if, for any $n \geq 1$ and $t_1, t_2, \cdots, t_n$,

$$\big(X_1(t_1), X_1(t_2), \cdots, X_1(t_n) \big) \stackrel{d}{=} \big(X_2(t_1), X_2(t_2), \cdots, X_2(t_n) \big),$$

where $\stackrel{d}{=}$ denotes equality of probability distributions. We will write this succinctly as $\{X_1(t)\} \stackrel{d}{=} \{X_2(t)\}$.

Lamperti (1962) introduced the idea of scaling in a process $X(t)$. He defined a d-dimensional process $X(t)$ as being *semi-stable* if it obeys a simple continuity condition and, for $s > 0$, the relationship

$$\{X(st)\} \stackrel{d}{=} \{b(s)X(t) + c(s)\}$$

holds, where $b(s)$ is a positive function and $c(s) \in \mathbb{R}^d$. Lamperti (1962) then showed that if $X(t)$ is a proper semi-stable process and $X(0) = 0$, then $c(s) = 0$ and $b(s) = s^H$ where H is a positive constant. That is,

$$\{X(st)\} \stackrel{d}{=} \{s^H X(t)\}, \tag{10.4}$$

which is now generally regarded as the definition of a *self-similar stochastic process*. Lamperti (1962) also showed that the normalising constant d_N in Equation 10.3 must be of the form $N^H L(N)$, where $L(\ )$ is a positive function that is slowly varying at infinity, and further, the limit process $X(t)$ must be effectively self-similar.

Taqqu (1975, 1977, 1978, 1979) in a series of papers considered a family of limiting self-similar stochastic processes originating from sums of the form given by Equation 10.3. Let $Y_j = h_m(Z_j)$ where $h_m(\ )$ is the Hermite polynomial of rank m and Z_j is a Gaussian sequence of random variables that have long range dependence whose autocorrelations are parameterised as a powerlaw function of H. Then it turns out that the limiting stochastic processes, $X(t)$, are self-similar. Further, when $m = 1$ the limiting processes is fractional Brownian motion, and when $m = 2$ it is the Rosenblatt process discussed above. Similar central limit type theorems were also introduced by Rosenblatt (1979, 1981), Dobrushin (1979), Dobrushin & Major (1979), and Major (1981). Taqqu & Wolpert (1983) discuss similar processes that are subordinated to a Poisson measure. See the text by Samorodnitsky & Taqqu (1994) for more details.

10.6.2 Self-Similar Stochastic Processes

The increments of a random function $\{X(t) : -\infty < t < \infty\}$ are said to be self-similar with parameter H if for any $s > 0$ and any τ

$$\{X(st + \tau) - X(\tau)\} \stackrel{d}{=} \{s^H(X(t + \tau) - X(\tau))\}.$$

If the increments of $X(t)$ are self-similar and $X(0) = 0$, then $X(t)$ is also self-similar as in Equation 10.4. If $X(t)$ has self-similar and stationary increments and is mean square continuous, then it can be shown that $0 \leq H < 1$.

The covariance structure can be derived directly from the above scaling law. Let $X(t)$ be a process with stationary self-similar increments. Then the covariance function is

$$\begin{aligned} \mathrm{E}[(X(t + \tau + 1) &- X(t + \tau))(X(t + 1) - X(t))] \\ &= \tfrac{1}{2}\sigma_H^2 \left\{|\tau + 1|^{2H} + |\tau - 1|^{2H} - 2|\tau|^{2H}\right\}, \end{aligned}$$

where $\sigma_H^2 = \mathrm{E}\big[(X(t + 1) - X(t))^2\big]$ for all t.

The process $X(t)$ is said to be *isotropic* if

$$\{X(t) - X(s)\} \stackrel{d}{=} \{X(|t - s|)\}. \qquad (10.5)$$

In this case, $\mathrm{E}\big[X^2(t)\big] = |t|^{2H}\mathrm{E}\big[X^2(1)\big]$, and

$$\mathrm{E}[X(t)X(s)] = \tfrac{1}{2}\left\{|s|^{2H} + |t|^{2H} - |t - s|^{2H}\right\}\mathrm{E}\big[X^2(1)\big].$$

10.6.3 Fractional Brownian Motion

A Gaussian process is uniquely determined by its autocovariance function. The unique Gaussian self-similar process is called *fractional Brownian motion*, which we will denote as $B_H(t)$. The increments of fractional Brownian motion are referred to as *fractional Gaussian noise*. If $B_H(0) = 0$, then the process $B_H(t)$ is isotropic (Equation 10.5).

Fractional Brownian Motion

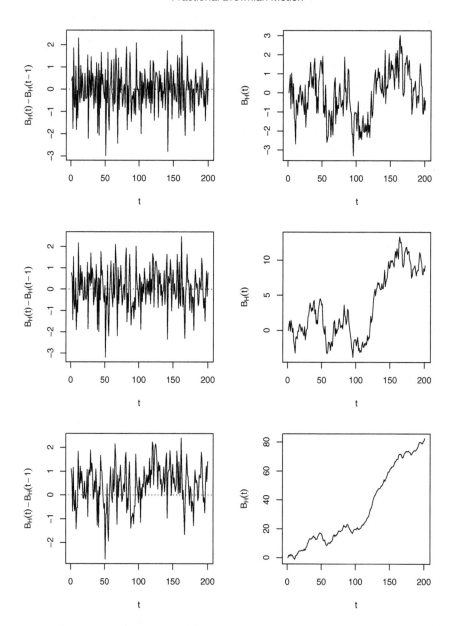

Figure 10.14 *Graphs of fractional Brownian motion (right) with* $H = 0.2, 0.5, 0.8$ *(top to bottom on right), with the increment processes on the left. The series have been simulated using the method of Davies & Harte (1987).*

When $H = \frac{1}{2}$, $B_H(t)$ is simply *Brownian motion*. When $H > \frac{1}{2}$ the autocorrelations are positive and have a powerlaw decay, hence long range dependence. When $H < \frac{1}{2}$ the correlations are negative and have a rapid decay. As $\tau \to \infty$

$$\mathrm{E}[(X(t + \tau + 1) - X(t + \tau))(X(t + 1) - X(t))] \approx H(2H - 1)\sigma^2 |\tau|^{2H-2},$$

(see Mandelbrot & Van Ness, 1968). These characteristics can be seen clearly in Figure 10.14, where the positive autocorrelation in the increment process can be seen for $H = 0.8$. Note also how the variance of $B_H(t)$ increases as H increases, consistent with the above relationships.

There are a number of results known about the sample path properties of fractional Brownian motion. It is a continuous process but non-differentiable. Orey (1970) shows that the Hausdorff dimension of the graph, i.e., the set of points $(t, B_H(t))$ on an interval with positive inner measure, is $2 - H$. Hence, when H is closer to zero, the graph (right-hand plots of Figure 10.14) will tend to be plane filling. Processes with long range dependence ($\frac{1}{2} < H < 1$) tend to be less space filling.

Figure 10.15 is a two dimensional plot of a sample path of fractional Brownian motion. The Hausdorff (and box) dimension of a path of Brownian motion in $\mathbb{R}^d$ is $\min(d, 2)$. Thus for $H = \frac{1}{2}$, we would expect the graphs to be space filling in localised regions. See Ciesielski & Taylor (1962) for $d \geq 3$ and Taylor (1964) for $d = 2$. For $H \neq 0.5$, Mandelbrot (1977, page 284, §7) suggests that the Hausdorff dimension of a path in $\mathbb{R}^d$ is $\min(d, 1/H)$. This is consistent with the graph for $H = 0.8$ which is less space filling. Related results for other auxiliary

2D Paths of Fractional Brownian Motion

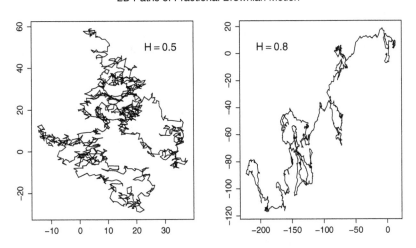

Figure 10.15 *Paths of two dimensional fractional Brownian motion with $H = 0.5$ and 0.8 for $t = 0, \ldots, 1500$. The series have been simulated using the method of Davies & Harte (1987).*

processes have also been given by Berman (1970) and Marcus (1976). Also see Adler (1981), Falconer (1990, Chapter 16) and Taylor (1986, page 390) for further discussion.

Davies & Hall (1999), Chan & Wood (2000), Constantine & Hall (1994) and Feuerverger et al. (1994) use the above ideas to characterise the roughness of a polished surface. The surface heights are measured at regular intervals, and the measurements are assumed to be sampled from a Gaussian like process with an autocorrelation function that is powerlaw. Orey (1970) also determined the Hausdorff dimension of the level crossings of a Gaussian process, and expressed it as a function of the powerlaw exponent found in the autocovariance function. Hence by estimating the powerlaw exponent in the autocovariance function, one can also calculate an estimate of the Hausdorff dimension of the level crossings. This can also be interpreted as a measure of surface roughness.

Davies & Harte (1987) describe test statistics for testing a hypothesis that a sampled time series is white noise against the alternative that it is fractional Gaussian noise.

10.6.4 Other 'Powerlaw' Processes

The fractionally differenced ARIMA process (see Hosking, 1981) also displays long range dependence. The fractional differencing is achieved by expressing the backward shift operator in the ARIMA model as an infinite binomial series expansion, that is

$$(1 - B)^d = \sum_{k=0}^{\infty} \binom{d}{k} (-B)^k.$$

Part of the motivation for the introduction of such a model was to provide the flexibility to model both the short term and long term correlation structure of an observed time series. Aspects of model fitting and building are discussed by Li & McLeod (1986). Geweke & Porter-Hudak (1983) compared the fractionally differenced ARIMA model to fractional Brownian motion, Haslett & Raftery (1989) used the process to model Ireland's wind power resource, and Porter-Hudak (1990) used it to model financial time series. Methods of estimation for the Hurst parameter H and the fractional differencing value d are discussed by Taqqu et al. (1995). The book by Beran (1994) provides a useful overview of long memory processes.

A process with $X(0) = 0$ whose increments are independent and belong to a stable law is also self-similar. Conversely, a process that is self-similar with independent increments must have increments that belong to a stable law (Lamperti, 1962). These processes are represented by their infinitely divisible characteristic function, see for example, Kolmogorov & Gnedenko (1954), Lukacs (1960), Fristedt (1974) and Breiman (1968). Fristedt (1974) also contains results on sample path properties.

10.6.5 'Multifractal' Stochastic Processes

More recently, some powerlaw stochastic processes have been described as being 'multifractal'. A multifractal stochastic process is defined by a moment function similar to $\theta(q)$, $\widetilde{\theta}(q)$, and $\theta^*(q)$ in Definitions 2.4.1, 2.3.1, and 6.1.2 respectively.

Let $\mathcal{T}$ and $\mathcal{Q}$ be intervals on the real line with positive length, such that $0 \in \mathcal{T}$ and $[0, 1] \subseteq \mathcal{Q}$. A stochastic process $X(t)$ is called multifractal if it has stationary increments and

$$\mathrm{E}[|X(t)|^q] = c(q)t^{\theta^\ddagger(q)+1}, \tag{10.6}$$

for all $t \in \mathcal{T}$ and all $q \in \mathcal{Q}$, where $c(q)$ and $\theta^\ddagger(q)$ are both function of $q \in \mathcal{Q}$ (see Mandelbrot et al., 1997).

Now consider a self-similar stochastic process that satisfies the scaling relation given by Equation 10.4. It follows that

$$\{X(t)\} \stackrel{d}{=} \{t^H X(1)\}$$

and hence

$$\mathrm{E}[|X(t)|^q] = t^{qH}\mathrm{E}[|X(1)|^q].$$

This satisfies the relation given by Equation 10.6 with $c(q) = \mathrm{E}[|X(1)|^q]$ and $\theta^\ddagger(q) = Hq - 1$. Note how $\theta^\ddagger(q)$ is a *linear* function of q. Recall that when this occurred in the multifractal measure setting, all Rényi dimensions were the same, and hence, while the measure may have been supported on a set with an extremely irregular shape, the allocation of measure within that set was in a sense uniform. These are sometimes referred to as *monofractal* measures. In this sense, a self-similar stochastic process is a monofractal process.

A stochastic process that satisfies a multiplicity of scaling laws will have a related $\theta^\ddagger(q)$ function that is non-linear. Assume that the process $X(t)$ satisfies the scaling relation

$$\{X(ct)\} \stackrel{d}{=} \{W(c)X(t)\} \tag{10.7}$$

for all t and $0 < c \leq 1$, where $W(c)$ is an independent stochastic process taking positive values; then

$$\mathrm{E}[|X(t)|^q] = \mathrm{E}[W(t)^q]\,\mathrm{E}[|X(1)|^q].$$

Further assume that $W(c)$ satisfies the following multiplicative property:

$$\{W(ab)\} \stackrel{d}{=} \{W_1(a)W_2(b)\},$$

where W_1 and W_2 are two independent copies of W, and a and b take values in the interval $(0, 1]$. It follows from the multiplicative property that for any t_1 and t_2 such that $t_1 t_2 = t$,

$$\mathrm{E}[W(t)^q] = \mathrm{E}[W(t_1)^q]\,\mathrm{E}[W(t_2)^q].$$

Hence $E[W(t)^q]$ must be of the form $t^{\theta^{\ddagger}(q)+1}$, and so $E[|X(t)|^q] = c(q)t^{\theta^{\ddagger}(q)+1}$, where $\theta^{\ddagger}(q)$ is a non-trivial function of q.

How does one define a stochastic process with multifractal behaviour? The method suggested by Mandelbrot et al. (1997) is to construct a compound process $X(t) = B_H(T(t))$, where $B_H(t)$ is fractional Brownian motion and $T(t)$ is a continuous non-decreasing random function of t that is independent of $B_H(t)$. That is, we have fractional Brownian motion in random time. In particular, let μ be a multifractal measure defined on the interval $[0, 1]$. Then one way to define $T(t)$ is as the cumulative distribution function of μ, i.e.,

$$T(t) = \frac{\mu([0, t])}{\mu([0, 1])} \ .$$

Mandelbrot (1999) outlines the sequence of historical models that have led to the development of this stochastic process. This process is also briefly discussed by Heyde (1999), who argues that some of its properties are unattractive for the modelling of financial data.

Our discussions on multifractal measures emphasized the relationships between a form of global averaging and local behaviour. In the above discussion, we have defined the multifractal process in terms of its global average. Its local or sample path behaviour is described by the scaling relation given by Equation 10.7. Riedi (2001) discusses the so called multifractal spectrum and formalism, and relates it to the moment functions discussed above.

Bouchaud et al. (2000) have analysed apparent multifractal behaviour in financial time series, and Fisher et al. (1997b) use these models to describe the Deutschemark–U.S. dollar exchange rates. Mandelbrot et al. (1997) and Fisher et al. (1997a) proposed these processes as models for financial returns. Riedi & Willinger (2000) suggest variations on these models to describe computer network traffic.

CHAPTER 11

Earthquake Analyses

11.1 Introduction

In this chapter, we estimate the Rényi dimensions using spatial point patterns produced by earthquakes and attempt to interpret these in an appropriate modelling context. These analyses follow on from the discussion started in §1.6.

Earthquakes are fractures in the earth's crust, however, they may not necessarily break the surface of the crust. The fracture may occur at a considerable depth, and there may be no obvious evidence on the surface of the earth. Earthquakes often occur near tectonic plate boundaries where there is considerable movement. In some situations, one boundary is undercutting its neighbour (plate subduction as in New Zealand or Japan), or the motion may be a horizontal slip (strike-slip as in much of California). There are also areas that have a considerable number of within plate earthquakes, for example China. Here the Eurasian plate is being squeezed by the subduction of the Pacific plate in Japan, and pressure being applied by the Indian plate in the area of the Himalayas (see Sphilhaus, 1991).

An earthquake is a fracture that starts at a point in time and space. However, the size of the fracture may extend over a considerable distance. The magnitude of the event is related to the size of the fracture. Wells & Coppersmith (1994) describe various relationships between the size of the rupture and the magnitude of the event. Earthquake catalogues contain the point in space (longitude, latitude and depth) at which the fracture commenced, the time of initiation, and the magnitude. The three-dimensional spatial location is referred to as the *hypocentre* and the two-dimensional surface location as the *epicentre*.

The intuitive motivation for estimating the fractal dimension of spatial point patterns generated by earthquakes is that the pattern may be self-similar in some sense. That is, clusters may be repeated within clusters on a finer and finer level. It is also thought that major fractures occur along major faults, the most dramatic being the tectonic plate boundaries. Within major fault systems there are smaller faults that branch off, and from these smaller fault networks; again with the possibility of generating some sort of self-similar hierarchy of networks. Seismicity is also highly clustered in time. However, a finite set of point locations theoretically has dimension zero, hence, what characteristics are the dimension estimates describing?

Dimension estimates are not a summary statistic of observed data in the same way as, for example, the sample mean. The dimension estimates are in fact describing a characteristic of the underlying probability distribution from which the data have been sampled. For example, the dimension estimates of the Moran cascade process of Example 10.3.1 described the Moran cascade probability measure, though the dimension estimates were based on a finite set of point observations. At this point one must ask: what is an appropriate modelling framework for earthquake events, and what is the interpretation of dimension estimates within such a framework? We use a point process framework which we briefly outline below.

11.1.1 Point Process Perspective

The forecasting of earthquakes is an extremely complex problem which is very much in its infancy. It is quite dissimilar to weather forecasting. Most earthquakes are not 'observable', only the effects of such events are observed if they are sufficiently large. Further, the estimated locations of events often contain considerable error, and the geophysical models describing the fracturing process are fairly primitive.

Unfortunately in the geophysics literature, there appears to be a dichotomy between 'geophysical' and 'statistical' models. The paradigm that we prefer is that there should be one model that includes the known geophysical aspects of the system, and describes the remaining stochastic aspects, that are not described by the geophysical components, with a statistical component.

One framework that can be used to model earthquake sequences is the point process model (Daley & Vere-Jones, 1988). A Poisson point process indexed by time can be characterised by its conditional intensity function. Let $N([t, t + \tau))$ be the number of events in $[t, t + \tau)$, and $\mathcal{H}_t$ be the history of the process up to but not including t. The conditional intensity function is defined as

$$\lambda(t|\mathcal{H}_t) = \lim_{\tau \to 0} \frac{1}{\tau} \Pr\{N([t, t + \tau)) > 0|\mathcal{H}_t\}. \tag{11.1}$$

It then follows that the expected number of events in an interval A, or the intensity measure, is

$$E[N(A)] = \int_A \lambda(t|\mathcal{H}_t)\, dt.$$

The process is naturally ordered by t. The above definition can be extended to also take the spatial aspect into account. Typically $\lambda(t|\mathcal{H}_t)$ will have a functional form that attempts to describe required geophysical characteristics of the process and other stochastic properties. It will also contain parameters that need to be estimated. This can be achieved by maximising the log-likelihood, which can be shown to be

$$\sum_{i=1}^{N} \log \lambda(t_i|\mathcal{H}_t) - \int \lambda(t_i|\mathcal{H}_t)\, dt,$$

where N is the number of observed events with occurrence times denoted by t_i, where $i = 1, \cdots, N$.

Note that the above formulation of a point process assumes that the events occur at a given point in time and space. If one wanted to include information describing the size and direction of the rupture, these could be included as a marks. See Daley & Vere-Jones (1988) for more information on marked point processes. As already noted, the magnitude is related to the size of the rupture, though such a scalar description may be too simple to describe both the size of the rupture and the amount of energy released, and a more useful representation may be gained by including tensor moments, etc.

Vere-Jones (1999) shows that there are at least two different interpretations of the Rényi dimension (D_q) estimates in the case of a space-time point process, where q is a positive integer ≥ 2. In the first, the spatial region is fixed and estimates are taken over increasing periods of time, and in the second, the time interval is fixed while the size of the spatial region increases. Vere-Jones (1999) then considers consistency results for both the correlation integral and correlation dimension estimates for both of the above scenarios.

In the first case where the spatial region is fixed: if the process is stationary and ergodic in time, and the cumulants satisfy certain regularity conditions, then Vere-Jones (1999) has shown that the Rényi dimension estimates are consistent for the Rényi dimensions associated with the spatial intensity measure of the process.

In the second where the observed period of time is fixed: if the process is stationary and ergodic in space, then Vere-Jones (1999) has shown that the Rényi dimension estimates are consistent for the initial powerlaw growth of the moment measures of the Palm distributions.

In this chapter, the Rényi dimensions of earthquake spatial locations are estimated and discussed from the perspective of a point process modelling framework. The analyses in this chapter are based on earlier work by Harte (1998).

11.2 Sources of Data

11.2.1 Determination of Earthquake Locations

When an earthquake occurs, the sudden release of an enormous amount of energy creates a P (primary) and S (secondary) wave. The P wave is a longitudinal wave, and the S wave is transverse. The P wave travels at a faster speed than the S wave. These waves will be detected by seismic stations, and from the trace that is plotted by the seismograph at each station, the arrival times, amplitudes, and duration of the P and S waves can be determined. By using a 'velocity model' which describes the speeds that the P and S waves should travel in the type of geological structures found in the vicinity of the earthquake, the earthquake location (rupture initiation point) and time can be determined by using the seismic trace from three or more seismic stations that recorded the event. The location algorithm uses an iterative least squares reweighting type procedure, adjusting the origin to minimise the residuals. Some earthquake catalogues contain the standard errors for each dimension from this location estimation procedure.

Geophysicists involved in earthquake research also use the travel times and other characteristics of seismic waves to determine the geological structure of the earth. It is this knowledge that can then be used to derive and modify velocity models. Some of the earthquake determination information, particularly the depth and magnitude are highly correlated. A velocity model that tends to overestimate the depth tends to underestimate the magnitude and visa versa. Effectively, the residuals from the fitting procedure contain a systematic component of error. These systematic errors can be substantial for events that occur some distance from seismic stations, for example, out to sea (see Harte & Vere-Jones, 1999). Hence the standard error may not necessarily give an accurate measure of the absolute error in the estimated earthquake location. Let x be the estimated

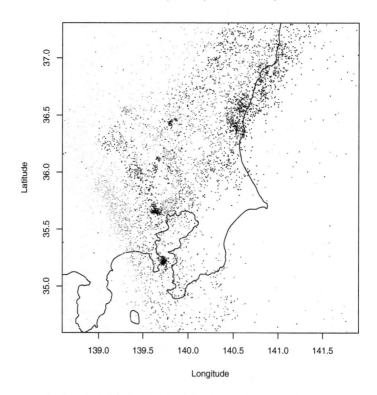

Kanto Earthquake Epicentres: Deep Events

Figure 11.1 *Kanto earthquake epicentres between 1985 and 1994, with magnitude ≥ 2 and depth $\geq 80\,km$. The deepest events are in the lightest shade of gray and the most shallow events are the darkest. The plot contains 7,570 events.*

hypocentre location of an event that occurs at x_0. Then

$$x = x_0 + S(x_0) + \epsilon,$$

where $S(x_0)$ is a systematic component of error, and ϵ is noise with zero mean. The standard errors will not describe the size of the systematic component, but simply $\text{Var}(\epsilon)$.

In our analyses we will use two earthquake catalogues, the Kanto Catalogue, and the Wellington Catalogue which was introduced in §1.6.1.

11.2.2 Kanto Earthquake Catalogue

The Kanto Earthquake Catalogue contains events from the Kanto region (Tokyo and surrounding area) of Japan. The catalogue is maintained by the National Research Institute for Earth Science and Disaster Prevention, Tsukuba-shi. For

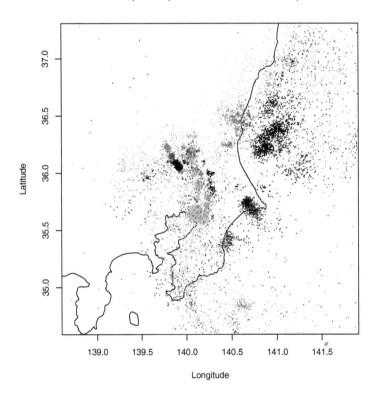

Figure 11.2 *Kanto earthquake epicentres between 1985 and 1994, with magnitude ≥ 2 and $40\,km \leq depth < 80\,km$. The deepest events are in the lightest shade of gray and the most shallow events are the darkest. The plot contains 12,486 events.*

this analysis, events have been selected with magnitude $\geq$ 2 between 138.6°E and 141.9°E, 34.6°N and 37.3°N, and occurring between 1 July 1979 and 30 June 1999.

The Kanto region is located on the junction of three tectonic plate boundaries, almost like a 'T' intersection (see Sphilhaus, 1991). To the west lies the Eurasian plate. The boundary of this plate runs roughly the length of Japan down toward Taiwan. In the Kanto region it is being subducted by the Pacific Plate to the east. To the southwest is the Philippine Plate which is also being subducted by the Pacific Plate. These boundaries are partially evident in Figures 11.1 and 11.2. In the lower left of both plots is a boundary of events running roughly from north-northwest toward south-southeast. This is the boundary between the Pacific and Philippines Plates. The other boundary in the top left of the picture, that runs roughly from the west-southwest to the east-northeast, marks the boundary

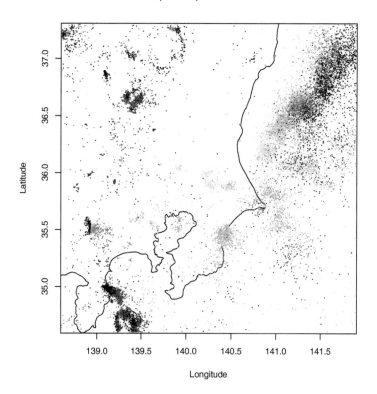

Figure 11.3 *Kanto earthquake epicentres between 1985 and 1994, with magnitude* $\geq$ 2 *and depth* < 40 km. *The deepest events are in the lightest shade of gray and the most shallow events are the darkest. The plot contains* 23,195 *events.*

between the Pacific and Eurasian Plates. In the Kanto region, the situation is more complicated than in the Wellington Region, and dense clusters of events tend to occur down to greater depths. However, it can be seen that most of the events with a depth $\geq$ 80 km are associated with the subduction process, whereas those more shallow events appear to have a more widespread distribution. We use 80 km as a boundary between intermediate and deep events for the Kanto Catalogue.

Figures 11.1, 11.2 and 11.3 are epicentral plots of deep, intermediate depth and shallow events respectively. Figure 11.4 shows the Pacific Plate subducting the Philippines Plate. Note that events in this subduction region outline the two friction boundaries of the subducting plate much more clearly than is the case in the Wellington Region (see Figure 1.6). This probably indicates less location error for events in the Kanto Catalogue. Also note, as in the Wellington Catalogue, that the shallow events tend to be much more clustered. Unlike the Wellington region, the Kanto region contains active volcanic zones, and a number of the events may be volcanic related, particularly off the Izu Peninsula.

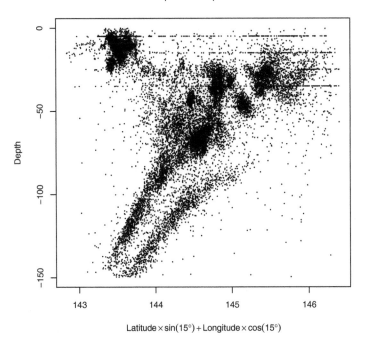

Figure 11.4 *Depth cross-section of Kanto earthquake locations between 1985 and 1994, with magnitude $\geq$ 2, depth $<$ 150 km, and south of 36.1°N. The plot contains 23,273 events. The picture shows the Pacific Plate subducting the Philippines Plate. The angle of view is (approximately) from the south-southeast to the north-northwest.*

11.3 Effects Causing Bias

11.3.1 Location Error

One of the most serious problems is event location error (i.e., noise in data). This causes the dimension estimates to rise to two for the epicentres and three for the hypocentres as the interpoint distance tends to zero, the same effect as depicted in Figure 9.3. It is not clear how one can correct for this, as the effect of adding noise effectively destroys any infinitesimal information contained in the data. An added complication in the present analyses is that this error is not constant over any analysed region. Tabulated standard errors in catalogues are a misleading measure of the hypocentre accuracy, in that they really describe the goodness of fit of the given earthquake velocity model and may not reflect systematic biases.

Papanastassiou & Matsumura (1987) studied errors of hypocentre locations in the Kanto Catalogue. At that stage the seismic network consisted of 67 stations: 17 surface stations, 2 tunnel stations, 45 shallow borehole stations, and 3 deep borehole stations. They were interested in determining the optimal location for a fourth deep borehole station. An update by Morandi & Matsumura (1991) reviewed the detection capability as a function of magnitude. Location errors tend to increase in the northeastern part of the analysed region. This is not only because of a lesser number of stations in this area (sea), but also due to the different geological structure. The western stations are more sensitive than the eastern ones because the western area is mountainous with low background noises due to the presence of hard bedrocks.

Papanastassiou & Matsumura (1987) provide contour plots for standard deviations of location errors for latitude, longitude, depth and the time of the event. These suggest error standard deviations in latitude to be less than 1.5 km for most of the region, increasing to 3 km in the east (off the coast) of the analysed region. Standard deviations in longitude are greater, though less than 2 km for most of the region, increasing to over 3 km in the east of the analysed region. Greatest errors occur in the depth variable, tending to increase for greater depth and again increasing to the east ($\approx$ 2–7 km).

A similar variation in the accuracy of earthquake locations can be expected in the Wellington Catalogue, with generally less accuracy in both offshore and deep events.

11.3.2 Boundaries and Lacunarity

Our method involves calculating the qth order interpoint distances between randomly selected events. Events within an ϵ distance of the imposed boundary of the region under study have a smaller chance of forming differences greater than ϵ, hence for larger values of m, $\widehat{\theta}_m$ is too small (Equation 8.8). This is referred to as the *boundary effect*. Bias is reduced by selecting one point in each pair from a restricted region $\mathcal{A}_\epsilon$ given by Equation 9.1. Further details are given in §9.2. This correction has not been implemented in the present analyses. While the boundaries used for the analyses define rectangular like regions, the actual geophysical boundaries within these analysed regions are much more irregular, and do not lend themselves to such a correction.

The Rényi dimensions are defined as a limiting operation, as the qth order interpoint distance (y) tends to zero. In practice, they are estimated using a section of the curve that is sufficiently flat. This is done under the assumption that if y is sufficiently small and $\theta(q)$ exists, the correlation integral $F_Y(y)$ will have power-law behaviour with exponent $\theta(q)$; i.e., there exists an ϵ such that for all $y < \epsilon$, $F_Y(y)$ is approximately powerlaw with exponent $\theta(q)$. However, if the boundary and location error effects are sufficiently severe, then their effects will merge, causing the line to be decreasing over all values of y and obscuring any underlying powerlaw behaviour.

Another type of boundary effect is caused by clusters of events. When the value of y is greater than the cluster diameter but smaller than the inter cluster distance, a deficit of sampled point differences occur, causing $\widehat{\theta}_m$ to decrease. It may be the case that clusters repeat themselves within clusters as in Figure 6.1. If the similarity ratios are the same, then such self-similar behaviour would exhibit itself as a *lacunary* cycle in the dimension estimates. If the similarity ratios are different, then the lacunary cycles will be out of phase with each other, and one will tend to cancel the effect of another. This will cause the dimension estimates to be flatter (see Figure 10.5).

11.3.3 Data Rounding and Transformation

An effect similar to lacunarity can also occur when data are rounded, restricting the interpoint distances to relatively few discrete points. We refer to this as the *rounding effect*, which was discussed in §9.3. In this situation, $\widehat{\theta}_m$ given by Equation 8.8 will also have an oscillatory behaviour, but unlike lacunarity, the period of the oscillation will be constant on the untransformed y scale. Another problem created by rounding is that *different* earthquake events are represented by the same spatial coordinates, causing problems with zero distances.

There are sometimes two levels of rounding. For example, in the case of the two analysed catalogues, most event depths are rounded to the closest one-tenth of a kilometre. However, in situations where shallow events occur at a considerable distance from any seismic recording station, the depth of the event will be poorly determined. In this situation, event depths in the Wellington Catalogue are restricted to 5, 12 or 33 km; and in the Kanto Catalogue, depths are restricted to 5, 15, 25 and 35 km.

Both catalogues had hypocentres tabulated in spherical coordinates, i.e., longitude and latitude as degrees east of Greenwich and south (or north) of the equator respectively, and depth in kilometres below the surface. Both longitude and latitude were transformed into a kilometre scale for these analyses. Hence all calculated interpoint distances always are in kilometre units. This may be at the expense of slight distortion in the extremities of the analysed regions.

11.4 Results

Estimates of the Rényi dimensions, $D_2, \cdots, D_5$ were made for events in different depth strata for both the Wellington and Kanto Catalogues. Wellington earthquakes were divided into two depth strata: shallow and deep, with events of depth < 40 km and depth ≥ 40 km respectively. Kanto earthquakes were divided into three depth strata: shallow, intermediate and deep, with events of depth < 40 km, 40 km $\leq$ depth < 80 km, and depth ≥ 80 km respectively. Dimension estimates for shallow and deep events from the Wellington Catalogue are plotted in Figures 11.5 and 11.6 respectively. Dimension estimates for shallow, intermediate and deep events from the Kanto Catalogue are plotted in Figures 11.7, 11.8, and 11.9 respectively.

When estimating dimensions, one hopes to find a relatively 'flat' interval in the plot over which the probability function of the interpoint distances scale in a powerlaw manner. Unfortunately, the effect of noise causes the line to increase to the dimension of the phase space for small decreasing interpoint distances, while the boundary effect causes the line to rapidly decrease for larger interpoint

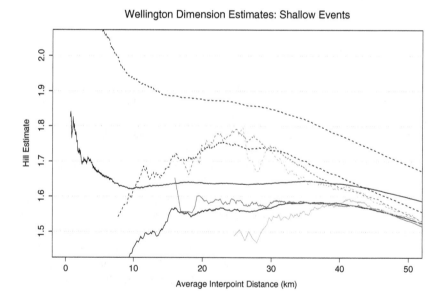

Figure 11.5 *Estimates of the Rényi dimensions D_2 (black), D_3 (dark gray), D_4 (light gray), and D_5 (lightest gray) for shallow events (depth < 40 km). The solid lines represent the estimates of the epicentres and the dashed lines of the hypocentres ($k = 100$, $n = 20{,}000$, $N = 15{,}228$).*

distances. When these two sources of bias are sufficiently severe, they merge and
no obvious flat interval exists. We have summarized our dimension estimates in
tables where the interval over which scaling has been estimated (or expected)
is tabulated, with the average estimate over that interval, and also the slope of
the line in that interval. A greater absolute value of the slope indicates that the
dimension estimate is more confounded with the noise and boundary sources of
bias, and hence is more uncertain. Thus, the slope gives a rough measure of the
extent to which the dimension estimate is affected by the noise and boundary
effects. While the line in some plots appears to be 'flat' or have a marked slope,
it is more a consequence of the scale of the graph, and hence the slope also gives
a slightly more objective measure with which to compare the different plots.

11.4.1 Wellington Catalogue

Estimates of Rényi dimensions for shallow events are plotted in Figure 11.5. The
dataset included those events with depth restricted to the values 0, 5, 12 and 33 km.
Note that out of the 15,228 shallow events, there were 942 events with depths
restricted to these values. Including or excluding these events did not make a noticeable difference. Using interpoint distances between 5 and 50 km, an estimate
of D_2 for epicentres is between 1.60 and 1.68, say 1.64. A flat region also appears

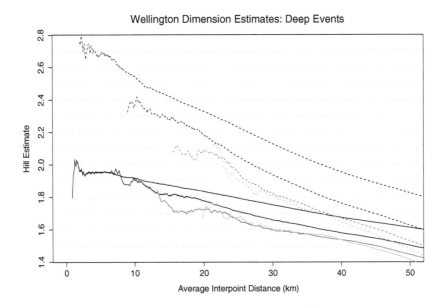

Figure 11.6 *Estimates of the Rényi dimensions D_2 (black), D_3 (dark gray), D_4 (light
gray), and D_5 (lightest gray) for deep events (depth $\geq$ 40 km). The solid lines represent
the estimates of the epicentres and the dashed lines of the hypocentres ($k = 100$, $n =
20,000$, $N = 6,746$).*

between 15 and 45 km in the case of D_3 and D_4 giving estimates of approximately 1.57. Estimates of D_5 are unstable, but appear to be similar to those of D_3 and D_4. Estimates based on hypocentres are less stable. This is not surprising as the depth of the region (40 km) is quite shallow relative to the width and breath, and hence the boundary effect would be expected to pose a greater problem. Further, the depth is probably the most poorly determined of the spatial coordinates, and hence the effect of noise extends further into the plot, affecting larger interpoint distances than is the case for the epicentres. As such we would expect to see a much smaller interval displaying powerlaw behaviour. Estimates and the scaling intervals have been summarized in Table 11.1.

Estimates of Rényi dimensions for deep events are plotted in Figure 11.6. A flat region is much less obvious here than for the shallow events. The deep events are mainly associated with the subducting slab which runs through the region roughly from southwest to the northeast as shown in Figure 1.6. The epicentres are the event locations when the hypocentre is projected onto the earth's surface. Consequently, the active epicentral area is relatively small, and hence the bound-

Wellington Catalogue			Scaling Range (km)	Estimate	| Slope |
Shallow	Epicentres	D_2	5–50	1.64	≈ 0
		D_3	15–45	1.57	≈ 0
		D_4	15–45	1.57	≈ 0
		D_5	35–45	1.57	unstable
	Hypocentres	D_2	15–30	1.87	≈ 0
		D_3	20–30	1.75	≈ 0
		D_4	20–30	1.75	≈ 0
		D_5		1.70	unstable
Deep	Epicentres	D_2	5–45	1.80	0.008
		D_3	5–45	1.70	0.008
		D_4	20–40	1.65	0.008
		D_5	20–40	1.65	0.008
	Hypocentres	D_2	5–50	2.25	0.020
		D_3	10–50	2.00	0.020
		D_4	10–50	1.80	0.020
		D_5	35–45	1.80	0.020

Table 11.1 *Summary of Rényi dimension estimates based on the Wellington Earthquake Catalogue. The 'slope' is the gradient of the line over the scaling range. Greater slopes indicate greater confounding with noise and boundary effects.*

ary effect will become a problem at relatively small interpoint distances. Even the hypocentre locations are contained within a relatively small region, one dimension being essentially the width of the subducting slab. There is probably also greater location error associated with deep events. With these two problems, it is not surprising that they have the effect of merging, causing the line of dimension estimates to decrease over the whole range of interpoint distances. We would expect powerlaw scaling to occur between about 5 and 35 km if the effects of these two biases were not as severe. Within this interval, the estimates of D_2 range between 1.95 and 1.68, giving a mid-point of about 1.8. Lines representing estimates of D_3 to D_5 are roughly parallel, with D_3 approximately 0.1 less than D_2 and both D_4 and D_5 0.15 less. Estimates are tabulated in Table 11.1.

Estimates representing the deep hypocentres decrease even more rapidly for increasing interpoint distance. There is really no 'flat' region, though this is partially due to the vertical scale of the plot. For each estimate of D_q, we have taken the mid point of the estimates in the interval of interpoint distances between 10 and 30 km. These values are tabulated in Table 11.1.

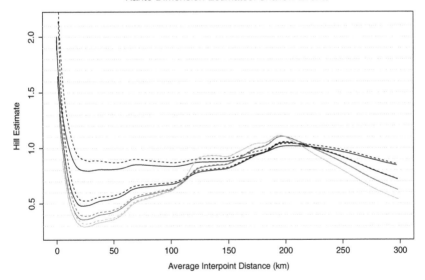

Figure 11.7 *Estimates of the Rényi dimensions D_2 (black), D_3 (dark gray), D_4 (light gray), and D_5 (lightest gray) for shallow events (depth $<$ 40 km). The solid lines represent the estimates of the epicentres and the dashed lines of the hypocentres ($k = 100$, $n = 40,000$, $N = 38,566$).*

11.4.2 Kanto Catalogue

Estimates of Rényi dimensions using shallow events (depth < 40 km) are plotted in Figure 11.7. Only estimates of D_2 show powerlaw behaviour. All estimates initially drop very rapidly then increase to a peak at approximately 200 km. For this interpoint distance, there is essentially no difference between the estimates of the epicentres and hypocentres. This is because 200 km is much greater than the total width of the depth dimension. While the difference between the dimension estimates of the epicentres and hypocentres appears to be very small, even for quite small interpoint distances, this is mainly due to the scale of the graph, and is roughly comparable to the Wellington shallow events in Figure 11.5. The peak in the dimension estimates at about 200 km is probably caused by this being the approximate distance between two or more very active clusters, which effectively dominate the characteristics of the whole plot. The dimension estimates of D_2 in Figure 11.7 also bear some resemblance to those of the beta distribution in Figure 10.6, which again is consistent with relatively few dominating clusters.

Estimates of Rényi dimensions using events of intermediate depth (40 km $\leq$ depth < 80 km) are plotted in Figure 11.8. Note that the lines representing epicentres tend to have a local maximum at approximately 60 km, whereas lines rep-

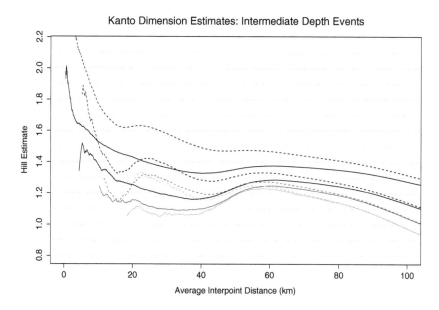

Figure 11.8 *Estimates of the Rényi dimensions D_2 (black), D_3 (dark gray), D_4 (light gray), and D_5 (lightest gray) for events of intermediate depth (40 km $\leq$ depth < 80 km). The solid lines represent the estimates of the epicentres and the dashed lines of the hypocentres ($k = 100$, $n = 25{,}000$, $N = 24{,}097$).*

Kanto Dimension Estimates: Deep Events

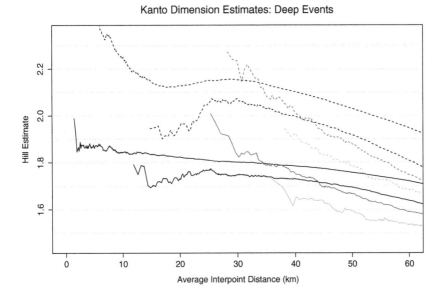

Figure 11.9 *Estimates of the Rényi dimensions D_2 (black), D_3 (dark gray), D_4 (light gray), and D_5 (lightest gray) for shallow events (depth $\geq$ 80 km). The solid lines represent the estimates of the epicentres and the dashed lines of the hypocentres ($k = 100$, $n = 15,000$, $N = 13,620$).*

resenting the hypocentres have two local maxima at approximately 25 and 60 km. These are probably caused by large clusters of events.

Estimates of Rényi dimensions for deep events (depth $\geq$ 80 km) are plotted in Figure 11.9. As for the deep events in the Wellington region, there is no interval where the lines are 'flat'. The estimates for the hypocentres of D_2 show a local minimum at slightly less than 20 km and a local maximum at approximately 30 km. This probably represents the distance between the two friction boundaries of the subducting slab, where the local maximum is caused by points in each pair being selected from the opposite boundary. Dimension estimates have been summarized in Table 11.2, giving the selected scaling interval, the average value of the dimension estimates, and the slope as for the Wellington Catalogue.

11.5 Comparison of Results and Conclusions

Dimension estimates for deep events in both regions are quite similar. Estimates in both regions are affect by the boundary and noise effects in a similar manner. This is indicated by the 'slope' in Tables 11.1 and 11.2 being similar. While the Kanto region contains active volcanic zones, the Wellington region does not. However, the deep events are those with depth $\geq$ 40 km and $\geq$ 80 km in the Wellington and Kanto regions, respectively. Volcanic induced seismicity occurs at shallower depths, and the analysed deep events will generally be associated with the plate subduction process in both regions.

The dimension estimates for shallow events in both regions are quite different. In the Wellington region, estimates display powerlaw scaling over a considerable range of distances. In the Kanto region, such scaling is only evident for D_2 and this does not commence until the interpoint distances are greater than 25 km. Since the maximum depth of shallow events is 40 km, one would expect this boundary effect to start to be noticeable at 25 km. This is why the dimension estimates for the hypocentres and epicentres are almost the same for interpoint distances greater than 25 km. Further, the estimates based on the Kanto Catalogue are considerably less than those based on the Wellington Catalogue. This probably

Kanto Catalogue			Scaling Range (km)	Estimate	| Slope |
Shallow	Epicentres	D_2	25–150	0.80	0.001
		D_3–D_5			uncertain
	Hypocentres	D_2	25–150	0.90	≈ 0
		D_3–D_5			uncertain
Intermediate	Epicentres	D_2	15–80	1.40	dips
		D_3	15–80	1.35	dips
		D_4	15–90	1.15	dips
		D_5	20–90	1.15	dips
	Hypocentres	D_2	15–45	1.60	uncertain
		D_3	15–45	1.55	uncertain
		D_4	15–45	1.35	uncertain
		D_5	20–45	1.35	uncertain
Deep	Epicentres	D_2	5–55	1.80	0.002
		D_3	15–45	1.75	0.002
		D_4	30–55	1.75	0.008
		D_5	40–60	1.60	0.008
	Hypocentres	D_2	15–35	2.15	dip
		D_3	15–40	2.00	unstable
		D_4	30–60	2.00	0.015
		D_5	40–60	1.80	0.015

Table 11.2 *Summary of Rényi dimension estimates based on the Kanto Earthquake Catalogue. The 'slope' is the gradient of the line over the scaling range. Greater slopes indicate greater confounding with noise and boundary effects.*

indicates that the spatial pattern of events in the Kanto region is characterised by clusters of greater density than in the Wellington region. This may be accounted for by the fact that the Wellington region is non-volcanic, while the Kanto region contains active volcanic areas. Earthquake clusters generated by volcanic activity tend to be both smaller and more dense. Clear estimates cannot be determined for D_3–D_5 for shallow events in the Kanto region.

Dimension estimates based on the Kanto Catalogue for events of intermediate depth are not monotonically decreasing (as interpoint distances increase) to the same extent as shown in other plots. They contain two local maxima at approximately 25 km and 60 km. Clear estimates cannot be determined for D_3–D_5 for hypocentres of intermediate depth in the Kanto region.

As the hypocentre depth increases, the estimated dimensions increase in both regions. This is consistent with the more uniform spatial patterns in Figures 1.8 and 11.1. This may partially be a consequence of location errors. These depend strongly on the angle between the active stations of the seismographic network as seen from a hypocentre; therefore we would expect that deep events would have larger location errors than the shallow events. This hypothesis is also consistent with the greater 'slope' in the plots of dimension estimates for the deep events compared to shallow events.

11.5.1 Point Process Setting

Lacunarity is caused by the support of a measure being a self-similar set with all similarity ratios being the same. This causes a periodic cycle in the dimension estimates when the interpoint distances are plotted on a logarithmic scale. When the similarity ratios are different, then the different induced lacunary cycles will be out of phase with each other and tend to cancel each other, producing a flatter curve of dimension estimates. A spatial intensity measure supported on a self-similar set with many different scaling ratios would produce similar dimension estimates to those of the shallow events in the Wellington region. Such an assumption of self-similarity seems quite reasonable, as this would relate to the self-similar nature of the stress distribution. Having different similarity ratios is also consistent with having geological heterogeneity within the region.

However, if one had an earthquake catalogue that included all events for many thousands of years, what values of dimension estimates would one expect? It appears from our analyses that given event clusters in space are also clustered in time. Thus viewing an epicentral plot in a given time period will produce noticeable clusters, and in another non-overlapping time interval will contain clusters which are non-existent in the first interval, and visa versa. It appears that 'hotspots' with intense activity in a given interval of time eventually die out and new 'hotspots' appear elsewhere. A related phenomenon occurs in volcanic zones where certain volcanos are active, then become extinct, and new volcanos emerge, though the time frame is much greater. It seems conceivable that if these 'hotspots' randomly appear, generating activity until stress is sufficiently relieved in that area, or possibly transferred to neighbouring areas (see Lu et al., 1999 and Bebbington & Harte, 2001), it could be argued that over a very long period of time, the fracturing process would tend to become space filling, and hence the Rényi dimensions of epicentres and hypocentres may tend to two and three respectively. If this argument is correct, then what is the interpretation of the dimension estimates in the point process context?

This argument relates most closely to the second case discussed by Vere-Jones (1999), where the spatial region is fixed, and as time increases indefinitely, the Rényi dimension estimates are consistent for those associated with the spatial intensity measure of the process. In the first case, the time period is held constant, and the spatial region is allowed to increase. This seems less plausible as the size of the region is essentially fixed and determined by the geophysical process.

The point process we propose is characterised by a conditional intensity function as in Equation 11.1, being conditional on the history of the process. Consider a model with a conditional intensity that is most dependent on the immediate past, and the current risk ultimately becomes independent of events in the distant past. As such the current stress distribution, and hence the current spatial intensity, could well be supported on a self-similar set, because it is not dependent on the 'complete' history. The spatial intensity measure may randomly evolve over time, though retain its self-similar character, as events in the distant past become independent of the current activity and hence are not influential on the current spatial stress distribution. One possibility when estimating Rényi dimensions is to sample events for the interpoint distances by weighting them according to their current influence which is specified by the given model.

If one had a catalogue containing data over many thousands of years, one could estimate the Rényi dimensions for the spatial intensity measure at various points in time, weighting the points according to their influence on the stress distribution for the time point being analysed. It would then be of interest to see if changes in multifractal characteristics of the spatial intensity measure over time is a precursor or coincides with detectable changes in observed seismicity.

In each situation analysed, estimates of the Rényi dimensions D_q decrease as q increases. If we were to accept that the spatial intensity measure of the process is supported on a self-similar set, then this decrease in the Rényi dimensions would further suggest that there is a greater likelihood of events occurring in certain parts of the region. If the spatial intensity measure is uniform, we would expect all Rényi dimensions to be the same.

11.5.2 Summary of Related Studies

There are a number of papers in the literature giving dimension estimates of earthquake locations. Hirata & Imoto (1991) estimate a range of Rényi dimensions of hypocentres in the Kanto region, getting a correlation dimension of about 2.2. Geilikman et al. (1990) also used a multifractal approach to investigate the spatial distribution of earthquake epicentres in Pamir, Caucasus and California. They estimated the Rényi dimensions using a box counting method, producing a box dimension, $\tilde{D}_0$, of about 2.0 for all regions.

Eneva (1996) estimated various Rényi dimensions for the spatial distribution of mining induced seismic activity. Her method was based on counting numbers of events within balls of certain radius, centred at points in the data set. She concluded that the apparent multifractal behaviour was spurious, and due to the limited size of the data sets.

Kagan (1991) used a worldwide catalogue of earthquakes to estimate the fractal dimension of epicentres and hypocentres. He found that as the time span of the catalogue increases his estimated value of the dimension of the hypocentres asymptotically reaches a value of 2.1–2.2 for shallow earthquakes. Further, his estimates of the fractal dimension decreased to 1.8–1.9 for intermediate depths (71–280 km) and to about 1.5–1.6 for deeper events. The decreasing dimension estimates for deeper events is the reverse of our results in this chapter.

Kagan's (1991) analyses *considerably larger* interpoint distances than we have done. His analyses are consistent with inspecting a worldwide epicentral map, where one would see that shallow earthquakes are concentrated in seismic belts, which extend more or less continuously and smoothly along the circum Pacific, and deep earthquakes are clustered in a few subduction zones. This suggests that the dimension estimates for deep events should be smaller than that for shallow events, at least for distances greater than about 50 km. We have analysed smaller localised regions, both of which are within subduction zones. The dimension of a set is defined as a limiting operation as distances tend to zero. However, if the set under consideration is self-similar, there may be powerlaw scaling over a considerable range of distances, with the powerlaw exponent being the same as the dimension. In order to use larger interpoint distances, two assumptions must be made: the legitimacy of the boundary adjustments, and that the powerlaw exponent at larger distances will be the same as that for smaller distances. We have analysed smaller interpoint distances, and as a result have had to contend with location errors of the hypocentre locations. The scale used for our analyses is roughly equivalent to the scales used in the modelling of regional seismicity.

Hypocentre location error may account for our dimension estimates being greater than Kagan's (1991) in the case of deep events, though this is not consistent with the estimates for epicentres. In the case of shallow events, we have achieved *smaller* values than Kagan (1991). This is inconsistent with the effect of location errors. This difference might be able to be explained by our analyses covering a relatively short time period. Kagan (1991) noted that dimension estimates tend to rise asymptotically to D_2 as the time span of the catalogue increases. However, both the Kanto and Wellington Catalogues have been analysed over almost the same length of time. Ogata & Abe (1991) suggest that seismic activity displays long term variation, therefore different spatial patterns may occur in times of higher seismic activity compared to lower activity.

PART IV

APPENDICES

Properties and Dimensions of Sets

In this appendix, definitions of the box, Hausdorff and packing dimensions are given, with various relationships between them. Most of the results closely follow the excellent book by Falconer (1990). We only provide a summary of some of the main results here, and much more detail can be found in Falconer (1990). Further mathematical detail can also be found in Falconer (1985). The book by Tricot (1995) also discusses various notions of dimension, but is more specific to curves.

A.1 Self-Similar Sets

A.1.1 Definitions (Falconer, 1990, Chapter 9)

1. Let E be a closed subset of $\mathbb{R}^d$. A mapping $S : E \to E$ is called a *contraction* on E if $\exists t : 0 < t < 1 : |S(x) - S(y)| \leq t|x - y|, \forall x, y \in E$.

2. Let E be a closed subset of $\mathbb{R}^d$. A mapping $S : E \to E$ is called a *similarity* on E if $\exists t : 0 < t < 1 : |S(x) - S(y)| = t|x - y|, \forall x, y \in E$. That is, S transforms sets to geometrically similar ones.

3. Let $S_1, \cdots, S_m$ be contractions. A subset F of E is called *invariant* for the transformations S_i if

$$F = \bigcup_{i=1}^{m} S_i(F). \tag{A.1}$$

4. A set that is invariant under a collection of similarities is called a *self-similar set*, that is, it is the union of a number of smaller copies of itself.

5. The contractions S_i are said to satisfy the *open set condition* if there exists a non-empty open bounded set V such that

$$V \supseteq \bigcup_{i=1}^{m} S_i(V) \tag{A.2}$$

with the union being disjoint. $\qquad\qquad\square$

Note the distinction between the open set condition and the disjoint set condition. In some papers, the open set condition may be described as *non-overlapping*; i.e., they only meet on the boundary points.

A.1.2 Theorem (Falconer, 1990, Page 114)

Let $S_1, \cdots, S_m$ be contractions on a subset E of $\mathbb{R}^d$ such that

$$|S_i(x) - S_i(y)| \leq t_i |x - y| \qquad x, y \in E,$$

with $t_i < 1$ for each i. Then there exists a *unique* non-empty *compact* set F that is invariant for the S_i s, i.e., that satisfies Equation A.1. Note that since $F \subseteq \mathbb{R}^d$ and is compact, then it must be closed.   □

A.1.3 Example

Let F be the middle third Cantor set on $[0, 1]$. Let $S_1(x) = x/3$ and $S_2(x) = x/3 + 2/3$. Then $F = S_1(F) \cup S_2(F)$, and so is self-similar. Further, let $V = (0, 1)$, then $S_1(V) = \left(0, \frac{1}{3}\right)$ and $S_2(V) = \left(\frac{2}{3}, 1\right)$, and so S_1 and S_2 satisfy the open set condition.   □

A.1.4 Definitions (Falconer, 1990, Chapter 9)

1. An *affine transformation* $A : \mathbb{R}^d \to \mathbb{R}^d$ is of the form $A(x) = T(x) + b$, where $x \in \mathbb{R}^d$ and T is a non-singular linear transformation.

2. Let $A_1, \cdots, A_m$ be affine contractions on $\mathbb{R}^d$. The set F which satisfies

$$F = \bigcup_{i=1}^{m} S_i(F).$$

   is called *self-affine*.   □

A.1.5 Theorem (Falconer, 1990, Page 118)

Suppose that the open set condition (Equation A.2) holds for similarities S_i on $\mathbb{R}^d$ with ratios t_i $(i = 1, \cdots, m)$. If F is the invariant set satisfying Equation A.1, then $\dim_H F = \dim_B F = s$, where s is given by

$$\sum_{i=1}^{m} t_i^s = 1.$$

Further, for this value of s, $0 < \mathcal{H}^s(F) < \infty$.   □

   Falconer (1990, §9.3) gives further results for the case where the S_i are contractions but not similarities.

A.2 Hausdorff Dimension

For a more detailed mathematical treatment see Falconer (1985, §1.2), Falconer (1990, §2.1) and Rogers (1970).

   Let U be a non-empty subset of $\mathbb{R}^d$. The diameter of U is defined as

$$|U| = \sup\{|x - y| : x, y \in U\}.$$

If $F \subset \bigcup_{i=1}^{\infty} U_i$ with $0 < |U_i| \leq \delta$ for each i, $\{U_i\}$ is called a δ-*cover* of F.

A.2.1 Hausdorff Measure Definition (Falconer, 1990, §2.1)

Suppose that F is a subset of $\mathbb{R}^d$ and s is a non-negative number. For any $\delta > 0$, define $\mathcal{H}_\delta^s(F)$ as

$$\mathcal{H}_\delta^s(F) = \inf \left\{ \sum_{i=1}^{\infty} |U_i|^s : \{U_i\} \text{ is a } \delta\text{-cover of } F \right\},$$

where the infimum is over all countable δ-covers $\{U_i\}$ of F. The s-*dimensional Hausdorff measure* of F is defined as

$$\mathcal{H}^s(F) = \lim_{\delta \to 0} \mathcal{H}_\delta^s(F).$$

□

The limit, $\mathcal{H}^s(F)$, exists for any $F \subset \mathbb{R}^d$, though the limiting value can often be zero or infinity.

A.2.2 Hausdorff Measure Properties (Falconer, 1990, §2.1)

We list below various properties of the Hausdorff measure.
1. (a) $\mathcal{H}^s(\emptyset) = 0$.
   (b) If $E \subseteq F$, then $\mathcal{H}^s(E) \leq \mathcal{H}^s(F)$.
   (c) If $\{F_i\}$ is any countable collection of disjoint Borel sets, then

$$\mathcal{H}^s \left(\bigcup_{i=1}^{\infty} F_i \right) = \sum_{i=1}^{\infty} \mathcal{H}^s(F_i).$$

2. If F is a Borel subset of $\mathbb{R}^d$, then

$$\mathcal{H}^d(F) = c_d \text{Vol}^d(F)$$

where $c_d = \pi^{d/2} 2^d (d/2)!$. For example, if F is a smooth surface, then

$$\mathcal{H}^2(F) = \frac{\pi}{4} \text{Area}(F).$$

3. *Scaling Property.* If $F \subset \mathbb{R}^d$ and $\lambda > 0$, then $\mathcal{H}^s(\lambda F) = \lambda^s \mathcal{H}^s(F)$ where $\lambda F = \{\lambda x : x \in F\}$.

4. Let $F \subset \mathbb{R}^d$ and $f : F \to \mathbb{R}^d$ be a mapping such that

$$|f(x) - f(y)| \leq c|x - y|^\alpha,$$

for $x, y \in F$, and for constants $c > 0$ and $\alpha > 0$. Then for each s,

$$\mathcal{H}^{s/\alpha}(f(F)) \leq c^{s/\alpha} \mathcal{H}^s(F).$$

□

If $t > s$ and $\{U_i\}$ is a δ-cover of F, then

$$\sum_{i=1}^{\infty} |U_i|^t \leq \delta^{t-s} \sum_{i=1}^{\infty} |U_i|^s,$$

so, taking the infimum over all possible coverings,

$$\mathcal{H}_{\delta}^t(F) \leq \delta^{t-s}\mathcal{H}_{\delta}^s(F).$$

Letting $\delta \to 0$, it can be seen that if $\mathcal{H}^s(F) < \infty$, then $\mathcal{H}^t(F) = 0$ for all $t > s$. Thus there is a critical value in the graph of $\mathcal{H}^s(F)$ against s at which it jumps from infinity to zero.

A.2.3 Hausdorff Dimension (Falconer, 1990, §2.2)

The Hausdorff dimension of a set F, denoted by $\dim_H(F)$, is

$$\dim_H(F) = \inf \{s : \mathcal{H}^s(F) = 0\} = \sup \{s : \mathcal{H}^s(F) = \infty\},$$

or alternatively satisfies

$$\mathcal{H}^s(F) = \begin{cases} \infty & \text{if } s < \dim_H(F) \\ 0 & \text{if } s > \dim_H(F) . \end{cases}$$

□

If $s = \dim_H(F)$, then $\mathcal{H}^s(F)$ may be zero, infinite, or may satisfy $0 < \mathcal{H}^s(F) < \infty$. A Borel set satisfying this last condition is called an s-set.

A.2.4 Properties of Hausdorff Dimension (Falconer, 1990, §2.2)

The Hausdorff dimension satisfies the following properties.

1. If $F \subset \mathbb{R}^d$ is open, then $\dim_H(F) = d$.

2. If F is a continuously differentiable m-dimensional submanifold of $\mathbb{R}^d$, then $\dim_H(F) = m$.

3. If $E \subset F$ then $\dim_H(E) \leq \dim_H(F)$.

4. If $F_1, F_2, \cdots$ is a countable sequence of sets then

$$\dim_H \left(\bigcup_{i=1}^{\infty} F_i \right) = \sup_{1 \leq i < \infty} \dim_H(F_i).$$

5. If F is countable then $\dim_H(F) = 0$.

6. If $F \subset \mathbb{R}^d$ is isometric with $E \subset \mathbb{R}^m$ $(m \leq d)$ then $\dim_H(F) = \dim_H(E)$. If E is of positive m-dimensional Lebesgue measure then

$$\dim_H(F) = \dim_H(E) = m.$$

□

A.2.5 Finer Definition (Falconer, 1990, §2.5)

Let $h : \mathbb{R}^+ \to \mathbb{R}^+$ be a continuous increasing function. Let

$$\mathcal{H}_\delta^h(F) = \inf \left\{ \sum_{i=1}^{\infty} h(|U_i|) : \{U_i\} \text{ is a } \delta\text{-cover of } F \right\},$$

for F a subset of $\mathbb{R}^d$. This leads to a measure by taking

$$\mathcal{H}^h(F) = \lim_{\delta \to 0} \mathcal{H}_\delta^h(F).$$

The function h is called the *dimension function*.   □

The definition of the s-dimensional Hausdorff measure has dimension function $h(t) = t^s$. If h and g are dimension functions such that

$$\frac{h(t)}{g(t)} \longrightarrow 0$$

as $t \to 0$ then it can be shown that $\mathcal{H}^h(F) = 0$ whenever $\mathcal{H}^g(F) < \infty$. For example, a set with dimension function $g(t) = t^s |\log t|$ would have Hausdorff dimension s, though the set is slightly smaller than one with dimension function $h(t)$.

S.J. Taylor has co-authored many papers on the dimension functions of various stochastic processes. A recent review is given by Taylor (1986).

A.3 Box Counting Dimension

Material in this section is covered in greater detail by Falconer (1990, §3.1).

A.3.1 Definition (Falconer, 1990, Page 38)

Let $N_\delta(F)$ be the smallest number of sets of diameter at most δ which cover the nonempty bounded set $F \subset \mathbb{R}^d$. The upper and lower box counting dimensions of F are given by

$$\underline{\dim}_B(F) = \liminf_{\delta \to 0} \frac{\log N_\delta(F)}{-\log \delta}$$

and

$$\overline{\dim}_B(F) = \limsup_{\delta \to 0} \frac{\log N_\delta(F)}{-\log \delta}$$

respectively. If $\underline{\dim}_B(F) = \overline{\dim}_B(F)$ then the *box counting dimension* of F is defined to be

$$\dim_B(F) = \lim_{\delta \to 0} \frac{\log N_\delta(F)}{-\log \delta}.$$

□

A.3.2 Theorem (Falconer, 1990, Page 40)

$N_\delta(F)$ can be any of the following. Each are equivalent.

1. The smallest number of closed balls of radius δ that cover F.
2. The smallest number of cubes of side δ that cover F.
3. The number of δ-mesh cubes of side δ that cover F.
4. The smallest number of sets of diameter at most δ that cover F.
5. The largest number of disjoint balls of radius δ with centers in F.   □

A.3.3 Note

$\underline{\dim}_B(F)$ may be defined by using economical coverings by small balls of equal radius (i.e., (1) above). $\overline{\dim}_B(F)$ may be thought of as a dimension that depends on coverings by disjoint balls of equal radius that are as dense as possible (i.e., (5) above). This concept forms the basis of the packing measure, defined later. □

A.3.4 Proposition (Falconer, 1990, Page 44)

Let $\overline{F}$ denote the closure of F (i.e., the smallest closed subset of $\mathbb{R}^d$ containing F). Then

$$\underline{\dim}_B(\overline{F}) = \underline{\dim}_B(F)$$

and

$$\overline{\dim}_B(\overline{F}) = \overline{\dim}_B(F).$$

□

A.3.5 Examples

The following examples demonstrate problems with the box counting dimension.

1. Let F be the (countable) set of rational numbers on $(0,1)$. Then $\overline{F} = [0,1]$ so that $\underline{\dim}_B(F) = \overline{\dim}_B(F) = 1$, and $\dim_B(F) = \dim_B(\overline{F}) = 1$. Thus countable sets can have a non-zero box counting dimension.

2. Let $F = \{0, 1, \frac{1}{2}, \frac{1}{3}, \cdots\}$, then F is a compact set with $\dim_B(F) = \frac{1}{2}$ (see Falconer, 1990, page 45).   □

A.3.6 Comparison with Hausdorff Dimension

Note that

$$N_\delta(F)\delta^s \longrightarrow \begin{cases} \infty & \text{if } s < \dim_B(F) \\ 0 & \text{if } s > \dim_B(F). \end{cases}$$

Now,

$$N_\delta(F)\delta^s = \inf\left\{\sum_i \delta^s : \{U_i\} \text{ is a finite } \delta\text{-cover of } F\right\},$$

and

$$\mathcal{H}_\delta^s(F) = \inf\left\{\sum_{i=1}^\infty |U_i|^s : \{U_i\} \text{ is a } \delta\text{-cover of } F\right\}.$$

As with the Hausdorff dimension, there may be a temptation to consider

$$\inf\left\{s : \liminf_{\delta\to 0} N_\delta(F)\delta^s = 0\right\}$$

as the box dimension. However, $\liminf_{\delta\to 0} N_\delta(F)\delta^s$ does not give a measure on subsets of $\mathbb{R}^d$. Thus, unlike the Hausdorff dimension, the box-counting dimension is not defined in terms of a measure. However, Pesin (1993, §2.2) defines upper and lower s-box measures.

The box counting dimension may be thought of as indicating the efficiency with which a set may be covered by small sets of *equal size*, whereas the Hausdorff dimension involves coverings by sets of small but possibly *widely varying size*. □

A.4 Packing Dimension

The packing dimension was introduced by Tricot (1982). It is defined in a similar way to the Hausdorff dimension, being based on a measure, though related in concept to $\overline{\dim}_B$ (see Note A.3.3). Let $F \subset \mathbb{R}^d$ and

$$P_\delta^s(F) = \sup\left\{\sum_{i=1}^\infty |B_i|^s : \begin{array}{l} \{B_i\}_\delta \text{ is a collection of disjoint balls of} \\ \text{radii at most } \delta \text{ with centres in } F \end{array}\right\}.$$

Since $P_\delta^s(F)$ decreases with δ, the limit

$$P_0^s(F) = \lim_{\delta\to 0} P_\delta^s(F)$$

exists. However, $P_0^s(F)$ is not a measure if F is a countable dense set. We therefore need a further modification.

A.4.1 Packing Measure Definition (Falconer, 1990, Page 47)

The *s-dimensional packing measure* of $F \subset \mathbb{R}^d$ is defined as

$$\mathcal{P}^s(F) = \inf\left\{\sum_{i=1}^\infty P_0^s(F_i) : F \subset \bigcup_{i=1}^\infty F_i\right\}.$$

□

A.4.2 Packing Dimension Definition (Falconer, 1990, Page 47)

The packing dimension, $\dim_P(F)$, is defined as

$$\dim_P(F) = \sup\left\{s : \mathcal{P}^s(F) = \infty\right\} = \inf\left\{s : \mathcal{P}^s(F) = 0\right\}.$$

☐

A.4.3 Note (Falconer, 1990, Page 47)

If $s > \dim_P(F_i)$ for all i, then

$$\mathcal{P}^s\left(\bigcup_{i=1}^{\infty} F_i\right) \leq \sum_{i=1}^{\infty} \mathcal{P}^s(F_i) = 0.$$

Thus $\dim_P\left(\bigcup_{i=1}^{\infty} F_i\right) \leq s$. Also note that

$$\dim_P\left(\bigcup_{i=1}^{\infty} F_i\right) = \sup_i \dim_P(F_i).$$

☐

A.4.4 Lemma (Falconer, 1990, Pages 43 & 48)

The following relationships hold for $F \subset \mathbb{R}^d$:

$$\dim_H(F) \leq \dim_P(F) \leq \overline{\dim}_B(F),$$

and

$$\dim_H(F) \leq \underline{\dim}_B(F) \leq \overline{\dim}_B(F).$$

☐

APPENDIX B

Large Deviations

B.1 Introduction

In this chapter we give a brief summary of results relating to large deviations that are referred to in Chapters 4, 5 and 6; particularly the Gärtner-Ellis Theorem. Most of the definitions and results have been taken from the book by Ellis (1985), where a considerably greater level depth can be found. The book by Bucklew (1990) provides a more accessible introduction to the subject.

Consider a sequence of random variables $Y_1, Y_2, \cdots$ converging in probability to a real constant y_0. That is,

$$\Pr\{|Y_n - y_0| > \epsilon\} \to 0 \quad \text{as} \quad n \to \infty.$$

It is often the case that $\Pr\{|Y_n - y_0| > \epsilon\}$ not only converges to zero, but does so exponentially fast. We could describe this convergence more explicitly by considering an ϵ interval about y, $(y - \epsilon, y + \epsilon)$, that does not necessarily contain the limit point y_0; and then investigate the probability $\Pr\{Y_n \in (y - \epsilon, y + \epsilon)\}$ as $n \to \infty$. It is often the case that

$$\Pr\{Y_n \in (y - \epsilon, y + \epsilon)\} \approx L(\epsilon, y, n) \exp[-nI_0(y, \epsilon)],$$

where $L(\epsilon, y, n)$ is non-negative, $\log L(\epsilon, y, n) = o(n)$ as $n \to \infty$, and $I_0(y, \epsilon)$ is a non-negative function that is zero when $y = y_0$. For a fixed $\epsilon > 0$ and a large index n, if $|Y_n - y_0| > \epsilon$, then Y_n can be thought of as a *large deviation* from the nominal value y_0, hence the terminology. Now consider re-expressing the above as

$$\frac{\log \Pr\{Y_n \in (y - \epsilon, y + \epsilon)\}}{n} \approx \frac{\log L(\epsilon, y, n)}{n} - I_0(y, \epsilon).$$

Since $\log L(\epsilon, y, n) = o(n)$ as $n \to \infty$, then for a fixed ϵ and sufficiently large n

$$\frac{\log \Pr\{Y_n \in (y - \epsilon, y + \epsilon)\}}{n} \approx -I_0(y, \epsilon).$$

Under fairly general conditions, this can be expressed as

$$\lim_{n \to \infty} \frac{1}{a_n} \log \Pr\{Y_n \in A\} = -\inf_{y \in A} I(y), \tag{B.1}$$

where $\{a_n; n = 1, 2, \cdots\}$ is a sequence of positive numbers tending to infinity, $A \in \mathcal{B}(\mathbb{R})$ and $I(y)$ is a non-negative convex function such that $I(y_0) = 0$. Much of the theory of large deviations is concerned with determining conditions under which this or similar statements can be made, and the form of the function $I(y)$.

Cramér's theorem considers Y_n to be a sequence of empirical means of independent identically distributed random variables, and shows that in this case the function $I(y)$ is related to the cumulant generating function. Cramér's theorem is reviewed in the next section.

Ellis (1984, 1985) and Gärtner (1977) consider a more general situation where the Y_n's take values in $\mathbb{R}^k$, are not necessarily empirical means, and may be defined on a sequence of probability spaces $(\Omega_n, \mathcal{F}_n, \rho_n)$ that are not necessarily the same. In the case where the Y_n's are defined on the same probability space $(\Omega, \mathcal{F}, \rho)$ and the sequence a_n, as in Equation B.1, increases sufficiently rapidly, then $Y_n \to y_0$ ρ-a.s. These results will be reviewed in the third section.

B.2 Cramér's Theorem

One of the earliest results is due to Harald Cramér. His result relies on the cumulant generating function being *steep*, which we now define.

B.2.1 Definition (Ellis, 1984)

Denote the *domain* of a function f as $\mathcal{D}(f) = \{x : f(x) < \infty\}$. A function $f : \mathbb{R}^k \to \mathbb{R}$ that is differentiable on the interior of $\mathcal{D}(f)$ is *steep* if, for all sequences $\{x_n\} \subseteq \mathcal{D}(f)$ that tend to a boundary point of $\mathcal{D}(f)$,

$$\|\nabla f(x_n)\| = \left\| \left(\frac{\partial f(x_n)}{\partial x_{n1}}, \cdots, \frac{\partial f(x_n)}{\partial x_{nd}} \right) \right\| \to \infty.$$

□

B.2.2 Cramér's Theorem (Bucklew, 1990, Page 7)

Let $\{X_i\}$, $i = 1, 2, \cdots$ be a sequence of independent and identically distributed random variables and

$$Y_n = \frac{1}{n} \sum_{i=1}^{n} X_i. \tag{B.2}$$

Suppose that $E[X_1]$ exists and is finite. Further, define the function $I(y)$ as

$$I(y) = \sup_q \{qy - C(q)\} \quad y \in \mathbb{R},$$

where $C(q)$ is the cumulant generating function of X_1, given by

$$C(q) = \log E\left[e^{qX_1}\right].$$

Assume that $C(q)$ is steep. Then for every closed set $K \subseteq \mathbb{R}$,

$$\limsup_{n\to\infty} \frac{1}{n} \log \Pr\{Y_n \in K\} \leq - \inf_{y \in K} I(y);$$

and for every open set $G \subseteq \mathbb{R}$,

$$\liminf_{n \to \infty} \frac{1}{n} \log \Pr\{Y_n \in G\} \geq - \inf_{y \in G} I(y).$$

□

B.2.3 Corollary

If, for $a, b \in \mathbb{R}$ and $a < b$,

$$\inf_{y \in [a,b]} I(y) = \inf_{y \in (a,b)} I(y),$$

then

$$\lim_{n \to \infty} \frac{1}{n} \log \Pr\{Y_n \in (a,b)\} = - \inf_{y \in (a,b)} I(y).$$

□

B.2.4 Example

Let $\{X_i\}$, $i = 1, 2, \cdots$ be a sequence of independent and identically distributed Gaussian random variables with mean μ and unit variance. Then the cumulant generating function is

$$C(q) = \log \mathrm{E}\left[e^{qX_1}\right] = \frac{q^2}{2} + \mu q \quad q \in \mathbb{R},$$

and so

$$I(y) = \frac{1}{2}(y - \mu)^2.$$

Since $q \in \mathbb{R}$ then there are no boundary problems with the domain of $C(q)$. Letting Y_n denote the empirical mean as in Equation B.2, then it follows from Cramér's theorem that

$$\lim_{n \to \infty} \frac{\log \Pr\{Y_n \in A\}}{n} = - \inf_{z \in A} I(z) \quad A \in \mathcal{B}(\mathbb{R}).$$

Consider various possibilities for the interval A.

1. Let $A = (\mu - \epsilon, \mu + \epsilon)$. Then $\inf_{z \in A} I(z) = 0$, thus

$$\lim_{n \to \infty} \frac{\log \Pr\{|Y_n - \mu| < \epsilon\}}{n} = 0.$$

2. Let $A = (-\infty, \mu - \epsilon) \cup (\mu + \epsilon, \infty)$. Then $\inf_{z \in A} I(z) = \epsilon^2/2$, thus

$$\lim_{n \to \infty} \frac{\log \Pr\{|Y_n - \mu| > \epsilon\}}{n} = -\frac{\epsilon^2}{2}.$$

Alternatively, $Y_n \sim \mathcal{N}(\mu, 1/n)$, therefore $\Pr\{|Y_n - \mu| > \epsilon\} = 2\Pr\{Z >$

$\sqrt{n}\epsilon\}$ where $Z \sim \mathcal{N}(0,1)$. It then follows from Abramowitz & Stegun (1964, Eq. 26.2.14) that, for $\epsilon > 0$

$$\Pr\{|Y_n - \mu| > \epsilon\}$$

$$= \left\{ \frac{1}{\sqrt{n}\epsilon+} \cdot \frac{1}{\sqrt{n}\epsilon+} \cdot \frac{2}{\sqrt{n}\epsilon+} \cdot \frac{3}{\sqrt{n}\epsilon+} \cdot \frac{4}{\sqrt{n}\epsilon+} \cdots \right\} \frac{2}{\sqrt{2\pi}} \exp\left(\frac{-n\epsilon^2}{2}\right).$$

The terms in the curly brackets are continued fractions. Taking logarithms of both sides, dividing by n and taking the limit in n gives the same result.

3. Now consider an ϵ interval about a point y, where $y \neq \mu$, i.e., $A = (y-\epsilon, y+\epsilon)$, and such that ϵ is sufficiently small so that $\mu \notin A$. Then

$$\inf_{z \in A} I(z) = \frac{1}{2}(|y - \mu| - \epsilon)^2,$$

thus

$$\lim_{n \to \infty} \frac{\log \Pr\{|Y_n - y| < \epsilon\}}{n} = -\frac{1}{2}(|y - \mu| - \epsilon)^2.$$

Taking limits in ϵ on both sides gives

$$\lim_{\epsilon \to 0} \lim_{n \to \infty} \frac{\log \Pr\{|Y_n - y| < \epsilon\}}{n} = -I(y).$$

This describes the *local* rate of convergence of the probability function.   □

B.2.5 Example

Let $\{X_i\}$, $i = 1, 2, \cdots$ be a sequence of independent and identically distributed exponential random variables with cumulant generating function

$$C(q) = \log \mathrm{E}\left[e^{qX_1}\right] = -\log\left(1 - \frac{q}{\lambda}\right) \qquad \lambda > 0 \text{ and } -\infty < q < \lambda,$$

thus $I(y) = \lambda y - 1 - \log(\lambda y)$. Note that the domain of $C(q)$ has a boundary at $q = \lambda$, however it can be seen that $C(q)$ is steep as $q \nearrow \lambda$. Letting Y_n denote the empirical mean as in Equation B.2, then it follows from Cramér's theorem that for $y > 1/\lambda$,

$$\lim_{n \to \infty} \frac{\log \Pr\{Y_n > y\}}{n} = -\inf_{z \in (y,\infty)} I(z) = 1 + \log(\lambda y) - \lambda y.$$

□

B.2.6 Example

Let $\{X_i\}$, $i = 1, 2, \cdots$ be a sequence of independent and identically distributed standard Gaussian random variables, and consider the partial sums Y_n, where

$$Y_n = \frac{1}{n} \sum_{i=1}^{n} X_i^2.$$

X_1^2 has a chi-squared distribution with cumulant generating function

$$C(q) = \log \mathrm{E}\left[e^{qX_1^2}\right] = \frac{-\log[1 - 2q]}{2} \qquad -\infty < q < \frac{1}{2},$$

and thus $I(y) = \frac{1}{2}(y - \log y - 1)$. It can be seen that $C(q)$ is steep at the boundary $q = \frac{1}{2}$, and so it follows from Cramér's theorem that for $y > 1$,

$$\lim_{n \to \infty} \frac{\log \Pr\{Y_n > y\}}{n} = -\inf_{z \in (y, \infty)} I(z) = \frac{-y + \log y + 1}{2}.$$

$\square$

B.3 Gärtner-Ellis Theorem

In this section the results are generalised to multiple dimensions, non-i.i.d., more general scaling constants, and situations that do not necessarily involve partial sums. Results are taken from Ellis (1984, 1985), and are similar to those of Gärtner (1977).

B.3.1 Definitions (Ellis, 1985, Chapter VI)

1. The set $A \subseteq \mathbb{R}^k$ is *convex* if $\lambda y_1 + (1 - \lambda)y_2$ is in A for every y_1 and y_2 in A, and every $0 < \lambda < 1$.

2. Let f be an extended real valued function defined on $A \subseteq \mathbb{R}^k$. f is said to be *convex on A* if

   (a) A is convex,
   (b) $f(y)$ is finite for at least one $y \in A$,
   (c) f does not take the value $-\infty$, and
   (d) $f(\lambda y_1 + (1 - \lambda)y_2) \leq \lambda f(y_1) + (1 - \lambda)f(y_2)$ for every y_1 and y_2 in A, and every $0 < \lambda < 1$.

3. Let f be a convex function on $\mathbb{R}^k$. If y_2 is a point in $\mathbb{R}^k$, then a vector $z \in \mathbb{R}^k$ is called a *subgradient* of f at y_2 (see §B.3.2 for an explanation) if

   $$f(y_1) \geq f(y_2) + \langle z, y_1 - y_2 \rangle \qquad \text{for all } y_1 \in \mathbb{R}^k.$$

4. The *subdifferential* of f at y is defined to be the set

   $$\partial f(y) = \{z \in \mathbb{R}^k : z \text{ is a subgradient of } f \text{ at } y\}$$

$\square$

B.3.2 Note

The notion of a subgradient extends the concept of a derivative to points at which f is not differentiable. It follows from above that if f is convex

$$\lambda(f(y_1) - f(y_2)) \geq f(y_2 + \lambda(y_1 - y_2)) - f(y_2).$$

Then if f is differentiable at y_2, then as $\lambda \to 0^+$

$$\lambda(f(y_1) - f(y_2)) \geq \lambda\langle\nabla f(y_2), y_1 - y_2\rangle + o(\lambda\|y_1 - y_2\|).$$

Rearranging, and letting $\lambda \to 0$ gives

$$f(y_1) \geq f(y_2)) + \langle\nabla f(y_2), y_1 - y_2\rangle.$$

$\square$

B.3.3 Preliminaries

Let $\{U_n; \ n = 1, 2, \cdots\}$ be a sequence of random vectors which are defined on probability spaces $\{(\Omega_n, \mathcal{F}_n, \rho_n); \ n = 1, 2, \cdots\}$, that is

$$U_n : (\Omega_n, \mathcal{F}_n) \to (\mathbb{R}^k, \mathcal{B}(\mathbb{R}^k)),$$

where $\mathcal{B}(\mathbb{R}^k)$ are the Borel sets of $\mathbb{R}^k$. We also consider the rescaled sequence of random vectors $\{Y_n; \ n = 1, 2, \cdots\}$, where

$$Y_n = \frac{U_n}{a_n},$$

and $\{a_n; \ n = 1, 2, \cdots\}$ is a sequence of positive numbers that tends to ∞. Let $\{Q_n; \ n = 1, 2, \cdots\}$ be a sequence of probability measures on $\mathcal{B}(\mathbb{R}^k)$, such that

$$Q_n\{B\} = \rho_n\{\omega \in \Omega_n : Y_n(\omega) \in B\}$$

where $B \in \mathcal{B}(\mathbb{R}^k)$.

$\square$

B.3.4 Definition (Ellis, 1985, Page 35)

Let $\{Q_n; \ n = 1, 2, \cdots\}$ be a sequence of probability measures on $\mathcal{B}(\mathbb{R}^k)$. $\{Q_n\}$ is said to have a *large deviation property* if there exists a sequence of positive numbers $\{a_n; n = 1, 2, \cdots\}$ which tend to ∞ and a function $I(y)$ which maps $\mathbb{R}^k$ into $[0, \infty]$ such that the following hypotheses hold.

1. $I(y)$ is closed (lower semi-continuous, i.e., $y_n \to y \Rightarrow \liminf_{n\to\infty} I(y_n) \geq I(y)$) on $\mathbb{R}^k$.

2. $I(y)$ has compact level sets, i.e., $\{y \in \mathbb{R}^k : I(y) \leq b\}$ is compact for each $b \in \mathbb{R}$.

3. For each closed set K in $\mathbb{R}^k$,

$$\limsup_{n\to\infty} \frac{1}{a_n} \log Q_n\{K\} \leq - \inf_{y\in K} I(y). \tag{B.3}$$

4. For each open set G in $\mathbb{R}^k$,

$$\liminf_{n\to\infty} \frac{1}{a_n} \log Q_n\{G\} \geq - \inf_{y\in G} I(y). \tag{B.4}$$

$I(y)$ is called the *entropy function* of $\{Q_n\}$.

$\square$

B.3.5 Theorem (Ellis, 1985, Page 36)

If for a fixed sequence $\{a_n; \ n = 1, 2, \cdots\}$, $\{Q_n\}$ has a large deviation property with entropy functions I and J, then $I(y) = J(y)$ for all $y \in \mathbb{R}^k$ (i.e., unique-ness).   □

B.3.6 Theorem (Ellis, 1985, Page 37)

Let Q_n be the distribution of $Y_n = U_n/a_n$ on $\mathbb{R}^k$. Suppose Q_n has a large devi-ation property with constants $\{a_n; n = 1, 2, \cdots\}$ and entropy function $I(y)$. The Borel set A will be called an I-continuity set if

$$\inf_{y \in \mathrm{cl} A} I(y) = \inf_{y \in \mathrm{int} A} I(y).$$

If A is an I-continuity set, then

$$\lim_{n \to \infty} \frac{1}{a_n} \log Q_n\{A\} = - \inf_{y \in A} I(y).$$

□

B.3.7 Rescaled Cumulant Generating Function

Define the function $C_n(q)$ as

$$
\begin{aligned}
C_n(q) &= \frac{1}{a_n} \log E_n[\exp\langle q, U_n \rangle] \\
&= \frac{1}{a_n} \log \int_{\Omega_n} \exp\langle q, U_n(\omega) \rangle \rho_n(d\omega), \quad\quad (\text{B.5})
\end{aligned}
$$

where $\{a_n; \ n = 1, 2, \cdots\}$ is a sequence of positive real numbers tending to infinity and $q \in \mathbb{R}^k$. If $\lim_{n \to \infty} C_n(q)$ exists, then $C(q) = \lim_{n \to \infty} C_n(q)$ is called the *Rescaled Cumulant Generating Function* of $\{U_n\}$. Note that $C(q)$ is often also called the *Free Energy Function*.   □

B.3.8 Hypotheses (Ellis, 1984, Hypothesis II.1)

The following hypotheses are assumed to be true.

1. $C(q) = \lim_{n \to \infty} C_n(q)$ exists for all $q \in \mathbb{R}^k$, where we allow $+\infty$ both as a limit value and as an element in the sequence $C_n(q)$ (i.e., define $C(q) = \infty$ if $C_n(q) = \infty$ for all $n > n_0$, n_0 depending on q).

2. $\mathcal{D}(C)$, where $\mathcal{D}(C) = \{q \in \mathbb{R}^k : C(q) < \infty\}$, has a non-empty interior containing the point $q = 0$.

3. $C(q)$ is a closed convex function on $\mathbb{R}^k$.   □

B.3.9 Notes (Ellis, 1984)

The domain of $C(q)$, $\mathcal{D}(C)$, is non-empty since $C(0) = 0 < \infty$. The hypotheses require that the *interior* of $\mathcal{D}(C)$ is non-empty. Convexity of $C(q)$ follows from the convexity of $C_n(q)$. The requirement of closure; i.e., for each real b, $\{q \in \mathbb{R}^k :$ $C(q) \leq b\}$ is closed in $\mathbb{R}^k$; is equivalent to $C(q)$ being lower semi-continuous. If $C(q)$ is closed and $\mathcal{D}(C)$ is an open set, then $C(q)$ is steep (Definition B.2.1). $\square$

B.3.10 Gärtner-Ellis Theorem (Ellis, 1984, Theorem II.2)

Let Q_n be the distribution of $Y_n = U_n/a_n$ on $\mathbb{R}^k$. Also define the function $I(y)$ as the *Legendre-Fenchel Transform* of $C(q)$, i.e.,

$$I(y) = \sup_{q \in \mathbb{R}^k} \{\langle q, y \rangle - C(q)\} \qquad y \in \mathbb{R}^k. \tag{B.6}$$

Then, given Hypotheses B.3.8, the following hold.

1. $I(y)$ is convex, closed (lower semi-continuous), and non-negative. $I(y)$ has compact level sets and $\inf_{y \in \mathbb{R}^k} I(y) = 0$.

2. The upper deviation bound, Equation B.3, is valid.

3. Assume in addition that $C(q)$ is differentiable on the interior of the domain of $C(q)$, and $C(q)$ is steep (Definition B.2.1). Then the lower large deviation bound, Equation B.4, is valid. $\square$

B.3.11 Corollary (Ellis, 1985, Theorem II.4.1)

Let $X_1, X_2, \cdots$ be a sequence of i.i.d. random vectors taking values in $\mathbb{R}^k$. Let $U_n = \sum_{i=1}^{n} X_i$, and $C(q)$ be given by Equation B.5. Assume that $C(q)$ is finite for all $q \in \mathbb{R}^k$, and define $I(y)$ by Equation B.6. Then the following hold.

1. The distributions of $Y_n = U_n/n$ on $\mathbb{R}^k$, denoted by $\{Q_n\}$, have a large deviation property with $a_n = n$ and entropy function $I(y)$.

2. $I(y)$ is a convex function of y. It measures the discrepancy between y and $\mathbb{E}[X_1]$ in the sense that $I(y) \geq 0$ with equality if and only if $y = \mathbb{E}[X_1]$.

3. If the domain of X_1 is the finite set $\{x_1, x_2, \cdots, x_r\} \subseteq \mathbb{R}$ with $x_1 < x_2 < \cdots < x_r$, then $I(y)$ is finite and continuous for $y \in [x_1, x_r]$ and $I(y) = \infty$ for $y \notin [x_1, x_r]$. $\square$

B.3.12 Lemma (Ellis, 1985, Page 214)

A convex function f on $\mathbb{R}$ ($k = 1$) is closed iff it is continuous on $\mathcal{D}(f)$ including

1. the end points if they are in $\mathcal{D}(f)$, i.e., $\mathcal{D}(f)$ is closed, or

2. $f(x) \to \infty$ as x approaches any end point not in $\mathcal{D}(f)$, i.e., $\mathcal{D}(f)$ is open and f is steep. $\square$

B.3.13 Definition - Exponential Convergence (Ellis, 1985, Page 231)

Let y_0 be a point in $\mathbb{R}^k$. $Y_n = U_n/a_n$ is said to *converge exponentially* to y_0, written

$$Y_n \xrightarrow{\text{exp}} y_0,$$

if for any $\epsilon > 0$, there exists a number $N = N(\epsilon) > 0$ such that

$$\rho_n \{\omega : \|Y_n(\omega) - y_0\| \geq \epsilon\} \leq \exp(-a_n N)$$

for all sufficiently large n.   □

B.3.14 Theorem (Ellis, 1984, Theorem IV.1)

Given Hypotheses B.3.8 hold, then the following statements are equivalent.

1. $Y_n = U_n/a_n \xrightarrow{\text{exp}} y_0$.

2. $C(q)$ is differentiable at $q = 0$ and $\nabla C(0) = y_0$.

3. $I(y)$ attains its minimum over $\mathbb{R}^k$ at the unique point $y = y_0$.   □

B.3.15 Theorem (Ellis, 1985, Page 49)

Assume that the random vectors $\{U_n; \ n = 1, 2, \cdots\}$ are all defined on the same probability space $(\Omega, \mathcal{F}, \rho)$. If $\sum_{n=1}^{\infty} \exp(-a_n N) < \infty$ for all $N > 0$, then $Y_n = U_n/a_n \xrightarrow{\text{exp}} y_0$ implies that $Y_n = U_n/a_n \xrightarrow{\text{a.s.}} y_0$.   □

B.3.16 Example

Let $\{X_i\}, i = 1, 2, \cdots$ be independent normal random variables with mean μ and unit variance. Define U_n as the partial sum to n as

$$U_n = X_1 + \cdots + X_n.$$

Let the rescaling constant be $a_n = n$ and $Y_n = U_n/a_n$, then the rescaled cumulant generating function of U_n is

$$
\begin{aligned}
C_n(q) &= \frac{1}{a_n} \log \mathrm{E}_n \left[e^{q U_n}\right] \\
&= \frac{1}{n} \log \mathrm{E}\left[e^{q X_1}\right]^n \\
&= \frac{q^2}{2} + \mu q \qquad q \in \mathbb{R}.
\end{aligned}
$$

Therefore

$$C(q) = \lim_{n \to \infty} C_n(q) = \frac{q^2}{2} + \mu q.$$

The problem is now equivalent to Example B.2.4. From Theorem B.3.14 it follows that

$$Y_n = \frac{X_1 + \cdots + X_n}{n} \xrightarrow{\text{exp}} C'(0) = \mu \qquad \text{as } n \to \infty.$$

Thus we have a weak law of large numbers, with the added strength that the convergence is exponentially fast. Since

$$\sum_{n=1}^{\infty} e^{-nN} = \frac{e^{-N}}{1 - e^{-N}} < \infty \qquad \forall N > 0,$$

then a strong law of large numbers follows from Theorem B.3.15, that is,

$$Y_n = \frac{X_1 + \cdots + X_n}{n} \xrightarrow{\text{a.s.}} \mu \qquad \text{as } n \to \infty.$$

$\square$

B.3.17 Theorem (Ellis, 1984, Theorem V.1)

Assume that Hypotheses B.3.8 hold. Then the following hold.

1. $I(y)$ is a closed convex function on $\mathbb{R}^k$.
2. $\langle q, y \rangle \leq C(q) + I(y)$ for all $q \in \mathcal{D}(C)$, and for all $y \in \mathcal{D}(I)$.
3. $\langle q, y \rangle = C(q) + I(y)$ if and only if $y \in \partial C(q)$.
4. $y \in \partial C(q)$ if and only if $q \in \partial I(y)$.
5. $C(q) = \sup_{y \in \mathbb{R}^k} \{\langle q, y \rangle - I(y)\}$ for all $q \in \mathbb{R}^k$.
6. For each real b, $\{y : I(y) \leq b\}$ is a closed, bounded, convex subset of $\mathbb{R}^k$.
7. $\inf_{y \in \mathbb{R}^k} I(y) = 0$, and $I(y_0) = 0$ if and only if $y_0 \in \partial C(q)$, which is a non-empty, closed, bounded, convex subset of $\mathbb{R}^k$. $\square$

References

Abarbanel, H.D.I. (1995). *Analysis of Observed Chaotic Data.* Springer-Verlag, New York.

Abarbanel, H.D.I., Brown, R., Sidorowich, J.J. and Tsimring, L.S. (1993). The analysis of observed chaotic data in physical systems. *Rev. Mod. Phys.* **65**(4), 1331–1392.

Abramowitz, M. and Stegun, I.A. (1964). *Handbook of Mathematical Functions.* Dover, New York.

Adler, R.J. (1981). *The Geometry of Random Fields.* John Wiley & Sons, Chichester.

Aki, K. (1989). Ideal probabilistic earthquake prediction. *Tectonophysics* **169**, 197–198.

Arbeiter, M. (1991). Random recursive construction of self-similar fractal measures. The compact case. *Probab. Th. Rel. Fields* **88**, 497–520.

Bebbington, M. and Harte, D.S. (2001). On the statistics of the linked stress release model. *J. Appl. Prob.* **38A**, 186–198.

Beck, C. (1990). Upper and lower bounds on the Rényi dimensions and the uniformity of multifractals. *Physica D* **41**, 67–68.

Beran, J. (1994). *Statistics for Long Memory Processes.* Chapman & Hall, New York.

Berliner, L.M. (1992). Statistics, probability and chaos. *Statistical Sci.* **7**(1), 69–122.

Berman, S.M. (1970). Gaussian processes with stationary increments: local times and sample function properties. *Ann. Statist.* **41**(4), 1260–1272.

Billingsley, P. (1965). *Ergodic Theory and Information.* John Wiley & Sons, New York.

Bohr, T., Jensen, M.H., Paladin, G. and Vulpiani, A. (1998). *Dynamical Systems Approach to Turbulence. Cambridge Nonlinear Science Series.* Cambridge University Press, Cambridge.

Bouchaud, J-P., Potters, M. and Meyer, M. (2000). Apparent multifractality in financial time series. *European Physical J. B* **13**, 595–599.

Breiman, L. (1968). *Probability.* Addison Wesley, Reading, MA.

Brown, G., Michon, G. and Peyrière, J. (1992). On the multifractal analysis of measures. *J. Statist. Physics* **66**(3/4), 775–790.

Bucklew, J.A. (1990). *Large Deviation Techniques in Decision, Simulation and Estimation.* John Wiley & Sons, New York.

Casdagli, M., Eubank, S., Farmer, J.D. and Gibson, J. (1991). State space reconstruction in the presence of noise. *Physica D* **51**(1-3), 52–98.

Cawley, R. and Mauldin, R.D. (1992). Multifractal decompositions of Moran fractals. *Adv. Math.* **92**, 196–236.

Chan, G. and Wood, A.T.A. (2000). Increment-based estimators of fractal dimension for two-dimensional surface data. *Statistica Sinica* **10**, 343–376.

Chatterjee, S. and Yilmaz, M.R. (1992). Chaos, fractals and statistics (with discussion). *Statistical Sci.* **7**(1), 49–68.

Cheng, B. and Tong, H. (1994). Orthogonal projection, embedding dimension and sample size in chaotic time series from a statistical perspective. *Phil. Trans. Royal Soc. Lond. A* **348**(1688), 325–341.

Ciesielski, E. and Taylor, S.J. (1962). First passage times and sojourn times for Brownian motion in space and the exact Hausdorff measure of the sample path. *Trans. Amer. Math. Soc.* **103**, 434–450.

Constantine, A.G. and Hall, P. (1994). Characterizing surface smoothness via estimation of effective fractal dimension. *J. R. Statist. Soc. B* **56**(1), 97–113.

Cutler, C.D. (1986). The Hausdorff dimension distribution of finite measures in Euclidean space. *Canad. J. Math.* **38**(6), 1459–1484.

Cutler, C.D. (1991). Some results on the behaviour and estimation of the fractal dimensions of distributions on attractors. *J. Statist. Physics* **62**(3/4), 651–708.

Cutler, C.D. (1994). A theory of correlation dimension for stationary time series. *Phil. Trans. Royal Soc. Lond. A* **348**(1688), 343–355.

Cutler, C.D. (1995). Strong and weak duality principles for fractal dimension in Euclidean space. *Math. Proc. Camb. Phil. Soc.* **118**(3), 393–410.

Cutler, C.D. (1997). A general approach to predictive and fractal scaling dimensions in discrete-index time series. *Fields Instit. Comm.* **11**, 29–48.

Cutler, C.D. and Dawson, D.A. (1989). Estimation of dimension for spatially distributed data and related limit theorems. *J. Multivar. Anal.* **28**, 115–148.

Cutler, C.D. and Kaplan, D.T. (1997). *Nonlinear Dynamics and Time Series: Building a Bridge Between the Natural and Statistical Sciences. Fields Institute Communications Volume 11.* American Mathematical Society, Providence, RI.

Cvitanović, P. (1993). *Universality in Chaos. 2nd Edition.* Institute of Physics Publishing, Bristol.

Daley, D.J. and Vere-Jones, D. (1988). *An Introduction to the Theory of Point Processes.* Springer-Verlag, New York.

David, H.A. (1970). *Order Statistics.* John Wiley & Sons, New York.

Davies, R.B. (1980). The distribution of a linear combination of χ^2 random variables. Algorithm AS 155. *Appl. Statist.* **29**(3), 323–333.

Davies, R.B. and Harte, D.S. (1987). Tests for Hurst effect. *Biometrika* **74**(1), 95–101.

Davies, S. and Hall, P. (1999). Fractal analysis of surface roughness by using spatial data (with discussion). *J. R. Statist. Soc. B* **61**(1), 3–37.

Denker, M. and Keller, G. (1986). Rigorous statistical procedures for data from dynamical systems. *J. Statist. Physics* **44**(1/2), 67–93.

Dobrushin, R.L. (1979). Gaussian and their subordinated self-similar random generalized fields. *Ann. Prob.* **7**(1), 1–28.

Dobrushin, R.L. and Major, P. (1979). Noncentral limit theorems for nonlinear functionals of Gaussian fields. *Z. Wahrscheinlichkeitstheorie verw. Gebiete* **50**, 27–52.

Dvořák, I. and Klaschka, J. (1990). Modification of the Grassberger-Procaccia algorithm for estimating the correlation exponent of chaotic systems with high embedding dimension. *Physics Letters A* **145**(5), 225–231.

Eckmann, J.-P. and Ruelle, D. (1985). Ergodic theory of chaos and strange attractors. *Rev. Mod. Phys.* **57**(3), 617–656.

Eckmann, J.-P. and Ruelle, D. (1992). Fundamental limitations for estimating dimensions and Lyapunov exponents in dynamical systems. *Physica D* **56**(2/3), 185–187.

Edgar, G.A. and Mauldin, R.D. (1992). Multifractal decompositions of digraph recursive fractals. *Proc. London Math. Soc.* **65**, 604–628.

Efron, B. and Tibshirani, R.J. (1986). Bootstrap methods for standard errors, confidence intervals, and other measures of statistical accuracy. *Statistical Sci.* **1**(1), 54–77.

Efron, B. and Tibshirani, R.J. (1993). *An Introduction to the Bootstrap.* Chapman & Hall, New York.

Eggleston, H.G. (1949). The fractional dimension of a set defined by decimal properties. *Quart. J. Math. (Oxford)* **20**, 31–36.

Ellis, R.S. (1984). Large deviations for a general class of random vectors. *Ann. Prob.* **12**(1), 1–12.

Ellis, R.S. (1985). *Entropy, Large Deviations, and Statistical Mechanics.* Springer-Verlag, New York.

Embrechts, P., Klüppelberg, C. and Mikosch, T. (1997). *Modelling Extremal Events for Insurance and Finance.* Springer-Verlag, New York.

Eneva, M. (1996). Effect of limited data sets in evaluating the scaling properties of spatially distributed data: an example from mining induced seismic activity. *Geophys. J. Int.* **124**(3), 773–786.

Essex, C. and Nerenberg, M.A.H. (1991). Comments on 'Deterministic chaos: the science and the fiction' by D. Ruelle. *Proc. R. Soc. Lond. A* **435**(1894), 287–292.

Falconer, K.J. (1985). *The Geometry of Fractal Sets.* Cambridge University Press, Cambridge.

Falconer, K.J. (1986). Random fractals. *Math. Proc. Camb. Phil. Soc.* **100**(3), 559–582.

Falconer, K.J. (1990). *Fractal Geometry, Mathematical Foundations and Applications.* John Wiley & Sons, Chichester.

Falconer, K.J. (1994). The multifractal spectrum of statistically self-similar measures. *J. Theor. Probab.* **7**(3), 681–702.

Falconer, K.J. (1997). *Techniques in Fractal Geometry.* John Wiley & Sons, Chichester.

Feller, W. (1971). *An Introduction to Probability Theory and its Applications. Volume 2, 2nd Edition.* John Wiley & Sons, New York.

Feuerverger, A., Hall, P. and Wood, A.T.A. (1994). Estimation of fractal index and fractal dimension of a Gaussian process by counting the number of level crossings. *J. Time Series Anal.* **15**(6), 587–606.

Fisher, A., Calvet, L. and Mandelbrot, B.B. (1997a). Large deviations and the distribution of prices changes. Cowles Foundation Discussion Paper 1165. Yale University, New Haven.

Fisher, A., Calvet, L. and Mandelbrot, B.B. (1997b). Multifractality of Deutschemark / US Dollar exchange rates. Cowles Foundation Discussion Paper 1166. Yale University, New Haven.

Fraser, A.M. and Swinney, H.L. (1986). Independent coordinates for strange attractors from mutual information. *Physical Rev. A* **33**(2), 1134–1140.

Friedlander, S.K. and Topper, L. (1961). *Turbulence: Classic Papers on Statistical Theory.* Interscience Publishers, New York.

Frisch, U. (1991). From global scaling, à la Kolmogorov, to multifractals scaling in fully developed turbulence. *Proc. R. Soc. Lond. A* **434**(1890), 89–99.

Fristedt, B. (1974). Sample functions of stochastic processes with stationary independent increments. In: *Advances in Probability. Vol 3*, 241–393. (Edited by P. Ney and S.C. Port). Marcel Dekker, New York.

Gács, P. (1973). Hausdorff-dimension and probability distributions. *Period. Math. Hungarica* **3**(1/2), 59–71.

Gärtner, J. (1977). On large deviations from the invariant measure. *Theor. Prob. Applic.* **22**(1), 24-39.

Geilikman, M.B., Golubeva, T.V. and Pisarenko, V.F. (1990). Multifractal patterns of seismicity. *Earth & Plan. Sci. Let.* **99**, 127–132.

Geller, R.J. (1997). Earthquake prediction: a critical review. *Geophys. J. Int.* **131**(3), 425–450.

Geller, R.J., Jackson, D.D., Kagan, Y.Y. and Mulargia, F. (1997). Earthquakes cannot be predicted. *Science* **275**(5306), 1616–1617.

Geweke, J. and Porter-Hudak, S. (1983). The estimation and application of long memory time series models. *J. Time Series Anal.* **4**(4), 221–238.

Graf, S. (1987). Statistically self-similar fractals. *Probab. Th. Rel. Fields* **74**(3), 357–392.

Grassberger, P. (1983). Generalized dimensions of strange attractors. *Physics Letters* **97A**(6), 227–230.

Grassberger, P. and Procaccia, I. (1983a). Measuring the strangeness of strange attractors. *Physica D* **9**(1/2), 189–208.

Grassberger, P. and Procaccia, I. (1983b). Estimation of the Kolmogorov entropy from a chaotic signal. *Physical Rev. A* **28**(4), 2591–2593.

Grassberger, P. and Procaccia, I. (1983c). Characterisation of strange attractors. *Physical Rev. Letters* **50**(5), 346–349.

Grebogi, C., Ott, E. and Yorke, J.A. (1988). Roundoff-induced periodicity and the correlation dimension of chaotic attractors. *Physical Rev. A* **38**(7), 3688–3692.

Gupta, Y.K. and Waymire, E.C. (1990). Multiscaling properties of spatial rainfall and river flow distributions. *J. Geophys. Res.* **95**(D3), 1999–2009.

Gupta, Y.K. and Waymire, E.C. (1993). A statistical analysis of mesoscale rainfall as a random cascade. *J. Appl. Meteorology* **32**(2), 251–267.

Halsey, T.C., Jensen, M.H., Kadanoff, L.P., Procaccia, I. and Shraiman, B.I. (1986). Fractal measures and their singularities: the characterisation of strange sets. *Physical Rev. A* **33**(2), 1141–1151.

Hao, B.L. (1990). *Chaos II.* World Scientific Publishing Company, Singapore.

Harte, D.S. (1998). Dimension estimates of earthquake epicentres and hypocentres. *J. Nonlinear Sci.* **8**, 581–618.

Harte, D.S. and Vere-Jones, D. (1999). Differences in coverage between the PDE and New Zealand local earthquake catalogues. *NZ. J. Geology & Geophysics* **42**, 237–253.

Haslett, J. and Raftery, A.E. (1989). Space-time modelling with long-memory dependence: assessing Ireland's wind power resource. *Appl. Statist.* **38**(1), 1–50.

Hentschel, H.G.E. and Procaccia, I. (1983). The infinite number of generalized dimensions of fractals and strange attractors. *Physica D* **8**(3), 435–444.

Heyde, C.C. (1999). A risky asset model with strong dependence through fractal activity time. *J. Appl. Prob.* **36**, 1234–1239.

Hill, B.M. (1975). A simple general approach to inference about the tail of a distribution. *Ann. Statist.* **3**(5), 1163–1174.

Hirata, T. and Imoto, M. (1991). Multifractal analysis of spatial distribution of microearthquakes in the Kanto region. *Geophys. J. Int.* **107**(1), 155–162.

Holley, R. and Waymire, E.C. (1992). Multifractal dimensions and scaling exponents for strongly bounded random cascades. *Ann. Appl. Prob.* **2**(4), 819–845.

Hosking, J.R.M. (1981). Fractional differencing. *Biometrika* **68**(1), 165–176.

Hunt, J.C.R., Phillips, O.M. and Williams, D. (1991). Turbulence and stochastic processes: Kolmogorov's ideas 50 years on. Preface and contents to this special issue. *Proc. R. Soc. Lond. A* **434**(1890), 5–7.

Hurst, H.E. (1951). Long term storage capacity of reservoirs. *Trans. Amer. Soc. of Civil Engineers* **116**, 770–808.

Hutchinson, J.E. (1981). Fractals and self similarity. *Indiana Uni. Math. J.* **30**(5), 713–747.

Isham, V. (1993). Statistical aspects of chaos: a review. In: *Networks and Chaos - Statistical and Probabilistic Aspects*, 124–200. (Edited by O. Barndorff-Nielsen, J.L. Jensen and W.S. Kendall). Chapman & Hall, London.

Johnson, N.L., Kotz, S. and Balakrishnan, N. (1994). *Continuous Univariate Distributions. Volume 2. Second Edition.* John Wiley & Sons, New York.

Kagan, Y.Y. (1981a). Spatial distribution of earthquakes: the three point moment function. *Geophys. J. R. Astron. Soc.* **67**, 697–717.

Kagan, Y.Y. (1981b). Spatial distribution of earthquakes: the four point moment function. *Geophys. J. R. Astron. Soc.* **67**, 719–733.

Kagan, Y.Y. (1991). Fractal dimension of brittle fracture. *J. Nonlinear Sci.* **1**, 1–16.

Kagan, Y.Y. (1997). Are earthquakes predictable? *Geophys. J. Int.* **131**(3), 505–525.

Kagan, Y.Y. (1999). Is earthquake seismology a hard, quantitative science? *PAGEOPH* **155**, 233–258.

Kagan, Y.Y. and Knopoff, L. (1980). Spatial distribution of earthquakes: the two point correlation function. *Geophys. J. R. Astron. Soc.* **62**, 303–320.

Kahane, J.-P. and Peyrière, J. (1976). Sur certaines martingales de Benoit Mandelbrot. *Adv. Math.* **22**(2), 131–145.

Kennel, M.B. and Isabelle, S. (1992). Method to distinguish possible chaos from colored noise and to determine embedding parameters. *Physical Rev. A* **46**(6), 3111–3118.

Kolmogorov, A.N. (1941). The local structure of turbulence in incompressible viscous fluid for very large Reynolds numbers. *Comptes Rendus (Doklady) de l'Aacdemie des Sciences de l'U.R.S.S.* **30**, 301–305. Reprinted in: Friedlander & Topper (1961).

Kolmogorov, A.N. (1962). A refinement of previous hypotheses concerning the local structure of turbulence in a viscous incompressible fluid at high Reynolds number. *J. Fluid Mechanics* **13**(1), 82–85.

Kolmogorov, A.N. and Gnedenko, B.V. (1954). *Limit Distributions for Sums of Independent Random Variables. (Translated by K.L. Chung.)* Addison Wesley, Reading, MA.

Lamperti, J. (1962). Semi-stable stochastic processes. *Trans. Amer. Math. Soc.* **104**, 62–78.

Lawrance, A.J. and Kottegoda, N.T. (1977). Stochastic modelling of river flow time series. *J. R. Statist. Soc. A* **140**(1), 1–31.

Lay, T. and Wallace, T.C. (1995). *Modern Global Seismology.* Academic Press, San Diego.

Li, W.K. and McLeod, A.I. (1986). Fractional time series modelling. *Biometrika* **73**(1), 217–221.

Lloyd, E.H. (1981). Stochastic hydrology: An introduction to wet statistics for dry statisticians. *Comm. Statist. A* **10**(15), 1505–1522.

Lorenz, E.N. (1963). Deterministic nonperiodic flow. *J. Atmos. Sci.* **20**, 130–141.

Lorenz, E.N. (1993). *The Essence of Chaos.* UCL Press, London.

Lovejoy, S. and Schertzer, D. (1985). Generalized scale invariance in the atmosphere and fractal models of rain. *Water Resources Res.* **21**(8), 1233–1250.

Lovejoy, S. and Schertzer, D. (1990). Multifractals, universality classes and satellite and radar measurements of cloud and rain fields. *J. Geophys. Res.* **95**(D3), 2021–2034.

Lu, C.S., Harte, D.S. and Bebbington, M. (1999). A linked stress release model for historical Japanese earthquakes: coupling among major seismic regions. *Earth Planets Space* **51**, 907–916.

Lukacs, E. (1960). *Characteristic Functions.* Griffin, London.

Major, P. (1981). Limit theorems for nonlinear functionals of Gaussian sequences. *Z. Wahrscheinlichkeitstheorie verw. Gebiete* **57**, 129–158.

Mandelbrot, B.B. (1974). Intermittent turbulence in self-similar cascades: divergence of high moments and dimension of the carrier. *J. Fluid Mechanics* **62**(2), 331–358.

Mandelbrot, B.B. (1975). Limit theorems on the self normalised range for weakly and strongly dependent processes. *Z. Wahrscheinlichkeitstheorie verw. Gebiete* **31**, 271–285.

Mandelbrot, B.B. (1977). *Fractals: Form, Chance and Dimension.* Freeman, San Francisco.

Mandelbrot, B.B. (1983). *The Fractal Geometry of Nature.* Freeman, New York.

Mandelbrot, B.B. (1989). Multifractal measures, especially for the geophysicist. *PAGEOPH* **131**(1/2), 5–42.

Mandelbrot, B.B. (1990a). Negative fractal dimensions and multifractals. *Physica A* **163**, 306–315.

Mandelbrot, B.B. (1990b). New "anomalous" multiplicative multifractals: left sided $f(\alpha)$ and the modelling of DLA. *Physica A* **168**, 95–111.

Mandelbrot, B.B. (1991). Random multifractals: negative dimensions and the resulting limitations of the thermodynamic formalism. *Proc. R. Soc. Lond. A* **434**(1890), 79–88.

Mandelbrot, B.B. (1997). *Fractals and Scaling in Finance : Discontinuity, Concentration, Risk : Selecta Volume E.* Springer-Verlag, New York.

Mandelbrot, B.B. (1998). *Multifractals and $1/f$ Noise: Wild Self-Affinity in Physics (1963–1976): Selecta Volume N.* Springer-Verlag, New York.

Mandelbrot, B.B. (1999). Renormalization and fixed points in finance, since 1962. *Physica A* **263**(1), 477–487.

Mandelbrot, B.B., Evertsz, C.J.G. and Hayakawa, Y. (1990). Exactly self-similar left-sided multifractal measures. *Physical Rev. A* **42**(8), 4528–4536.

Mandelbrot, B.B., Fisher, A. and Calvet, L. (1997). A multifractal model of asset returns. Cowles Foundation Discussion Paper 1164. Yale University, New Haven.

Mandelbrot, B.B. and Riedi, R.H. (1997). Inverse measures, the inversion formula, and discontinuous multifractals. *Adv. Appl. Math.* **18**(1), 50–58.

Mandelbrot, B.B. and Van Ness, J.W. (1968). Fractional Brownian motions, fractional noises and applications. *SIAM Review* **10**(4), 422–437.

Mandelbrot, B.B. and Wallis, J.R. (1969). Robustness of the rescaled range R/S in the measurement of non-cyclic long run statistical dependence. *Water Resources Res.* **5**(5), 967–988.

Marcus, M.B. (1976). Capacity of level sets of certain stochastic processes. *Z. Wahrscheinlichkeitstheorie verw. Gebiete* **34**, 279–284.

Mason, D.M. (1982). Laws of large numbers for sums of extreme values. *Ann. Prob.* **10**(3), 754–764.

Mauldin, R.D. and Williams, S.C. (1988). Random recursive constructions: asymptotic geometric and topological properties. *Trans. Amer. Math. Soc.* **295**(1), 325–346.

Maunder, D.E. (1994). New Zealand Seismological Report 1992. Seismological Observatory Bulletin E-176. Institute of Geological and Nuclear Sciences Report 94/47. Institute of Geological and Nuclear Sciences, Lower Hutt.

May, R.M. (1976). Simple mathematical models with very complicated dynamics. *Nature* **261**, 459–467.

May, R.M. (1987). Chaos and the dynamics of biological populations. *Proc. R. Soc. Lond. A* **413**(1844), 27–44.

Meneveau, C. and Sreenivasan, K.R. (1991). The multifractal nature of turbulent energy dissipation. *J. Fluid Mechanics* **224**, 429–484.

Mikosch, T. and Wang, Q. (1993). Some results on estimating Rényi type dimensions. Institute of Statistics and Operations Research Report. Victoria University of Wellington, Wellington.

Mikosch, T. and Wang, Q. (1995). A Monte-Carlo method for estimating the correlation exponent. *J. Statist. Physics* **78**(3/4), 799–813.

Molchan, G.M. (1995). Multifractal analysis of Brownian zero set. *J. Statist. Physics* **79**(3/4), 701–730.

Monin, A.S. and Yaglom, A.M. (1971). *Statistical Fluid Mechanics; Mechanics of Turbulence.* MIT, Cambridge, MA.

Moran, P.A.P (1946). Additive functions of intervals and Hausdorff measure. *Proc. Camb. Phil. Soc.* **42**, 15–23.

Moran, P.A.P. (1950). Some remarks on animal population dynamics. *Biometrics* **6**(3), 250–258.

Morandi, M.T. and Matsumura, S. (1991). Update on the examination of the seismic observational network of the National Research Institute for Earth Science and Disaster Prevention (NIED): detection capability and magnitude correction. No. 47, 1–18. National Research Institute for Earth Science and Disaster Prevention, Tsukuba.

Nerenberg, M.A.H. and Essex, C. (1990). Correlation dimension and systematic geometric effects. *Physical Rev. A* **42**(12), 7065–7074.

Nicolis, C. and Nicolis, G. (1984). Is there a climate attractor? *Nature* **311**, 529–532.

Oboukhov, A.M. (1962). Some specific features of atmospheric turbulence. *J. Fluid Mechanics* **13**(1), 77–81.

Ogata, Y. and Abe, K. (1991). Some statistical features of the long term variation of the global and regional seismic activity. *Int. Statist. Rev.* **59**(2), 139–161.

Olsen, L. (1994). *Random Geometrically Graph Directed Self-Similar Multifractals.* Longman Scientific & Technical, Harlow, Essex.

Orey, S. (1970). Gaussian sample functions and the Hausdorff dimension of level crossings. *Z. Wahrscheinlichkeitstheorie verw. Gebiete* **15**, 249–256.

Osborne, A.R. and Provenzale, A. (1989). Finite correlation dimension for stochastic systems with power-law spectra. *Physica D* **35**(3), 357–381.

Ott, E., Sauer, T. and Yorke, J.A. (1994). *Coping with Chaos. Analysis of Chaotic Data with the Exploitation of Chaotic Systems.* John Wiley & Sons, New York.

Pachard, N.H., Crutchfield, J.P., Farmer, J.D. and Shaw, R.S. (1980). Geometry from a time series. *Physical Rev. Letters* **45**(9), 712–716.

Paladin, G. and Vulpiani, A. (1987). Anomalous scaling laws in multifractal objects. *Physics Reports* **156**(4), 147–225.

Papanastassiou, D. and Matsumura, S. (1987). Examination of the NRCDP's (The National Research Center for Disaster Prevention) seismic observational network as regards:

I. detectability-locatability, II. accuracy of the determination of the earthquake source parameters. No. 39, 37-65. National Research Institute for Earth Science and Disaster Prevention, Tsukuba.

Park, K. and Willinger, W. (Editors) (2000). *Self-Similar Network Traffic and Performance Evaluation.* John Wiley & Sons, New York.

Pesin, Y.B. (1993). On rigorous mathematical definitions of correlation dimension and generalized spectrum for dimensions. *J. Statist. Physics* **71**(3/4), 529–547.

Pisarenko, D.V. and Pisarenko, V.F. (1995). Statistical estimation of the correlation dimension. *Physics Letters A* **197**(1), 31–39.

Porter-Hudak, S. (1990). An application of the seasonal fractionally differenced model to the monetary aggregates. *J. Amer. Statist. Assoc.* **85**(410), 338–344.

Rasband, S.N. (1990). *Chaotic Dynamics of Nonlinear Systems.* John Wiley & Sons, New York.

Rényi, A. (1959). On the dimension and entropy of probability distributions. *Acta Mathematica* **10**, 193–215.

Rényi, A. (1965). On the foundations of information theory. *Review of I.S.I.* **33**(1), 1–14.

Rényi, A. (1970). *Probability Theory.* Elsevier Science Publishers B.V. (North-Holland), Amsterdam.

Resnick, S.I. (1997). Heavy tail modeling and teletraffic data. *Ann. Statist.* **25**(5), 1805–1869.

Riedi, R. (1995). An improved multifractal formalism and self-similar measures. *J. Math. Anal. Applic.* **189**(2), 462–490.

Riedi, R.H. (2001). Multifractal processes. In: *Long-Range Dependence: Theory and Applications.* (Edited by: by Murad S. Taqqu, George Oppenheim, Paul Doukhan). Birkhäuser, Boston.

Riedi, R.H. and Mandelbrot, B.B. (1995). Multifractal formalism for infinite multinomial measures. *Adv. Appl. Math.* **16**(2), 132–150.

Riedi, R.H. and Mandelbrot, B.B. (1997). Inversion formula for continuous multifractals. *Adv. Appl. Math.* **19**(3), 332–354.

Riedi, R.H. and Mandelbrot, B.B. (1998). Exceptions to the multifractal formalism for discontinuous measures. *Math. Proc. Camb. Phil. Soc.* **123**(1), 133–157.

Riedi, R.H. and Willinger, W. (2000). Toward an improved understanding of network traffic dynamics. In: *Self-Similar Network Traffic and Performance Evaluation.* (Edited by: Kihong Park and Walter Willinger). John Wiley & Sons, New York.

Rogers, C.A. (1970). *Hausdorff Measures.* Cambridge University Press, Cambridge.

Rosenblatt, M. (1961). Independence and dependence. In: *Proceedings of the 4th Berkeley Symposium on Mathematical Statistics and Probability. Volume II. Probability Theory,* 431–443. (Edited by J. Neyman). University of California Press, Berkley.

Rosenblatt, M. (1979). Some limit theorems for partial sums of quadratic forms in stationary Gaussian variables. *Z. Wahrscheinlichkeitstheorie verw. Gebiete* **49**, 125–132.

Rosenblatt, M. (1981). Limit theorems for Fourier transforms of functionals of Gaussian sequences. *Z. Wahrscheinlichkeitstheorie verw. Gebiete* **55**, 123–132.

Ruelle, D. (1989). *Chaotic Evolution and Strange Attractors.* Cambridge University Press, Cambridge.

Ruelle, D. (1990). The Claude Bernard Lecture, 1989. Deterministic chaos: the science and the fiction. *Proc. R. Soc. Lond. A* **427**(1873), 241–248.

Ruelle, D. (1995). *Turbulence, Strange Attractors, and Chaos.* World Scientific Publishing Company, Singapore.

Samorodnitsky, G. and Taqqu, M.S. (1994). *Stable Non-Gaussian Random Processes: Stochastic Models with Infinite Variance.* Chapman & Hall, New York.

Sauer, T., Yorke, J.A. and Casdagli, M. (1991). Embedology. *J. Statist. Physics* **65**(3/4), 579–616.

Schertzer, D. and Lovejoy, S. (1987). Physical modeling and analysis of rain and clouds by anisotropic scaling multiplicative processes. *J. Geophys. Res.* **92**(D8), 9693–9714.

Schertzer, D. and Lovejoy, S. (1989). Generalised scale invariance and multiplicative processes in the atmosphere. *PAGEOPH* **130**(1), 57–81.

Scholz, C.H. (1990). *The Mechanics of Earthquakes and Faulting.* Cambridge University Press, Cambridge.

Scholz, C.H. and Mandelbrot, B.B. (1989). *Fractals in Geophysics. Reprint from PAGEOPH 131(1/2).* Birkhäuser Verlag, Basel.

Smith, R.L. (1992a). Optimal estimation of fractal dimension. In: *Nonlinear Modeling and Forecasting, SFI Studies in the Sciences of Complexity. Proceedings of the Workshop on Nonlinear Modeling and Forecasting, September 1990, at Santa Fe, New Mexico. Vol XII,* 115–135. (Edited by M. Casdagli and S. Eubank). Addison Wesley, Redwood City, CA.

Smith, R.L. (1992b). Estimating dimension in noisy chaotic time series. *J. R. Statist. Soc. B* **54**(2), 329–351.

Sphilhaus, A. (1991). *Atlas of the World and Geophysical Boundaries Showing Oceans, Continents and Tectonic Plates in Their Entirety.* American Philosophical Society, Philadelphia.

Stoyan, D. and Stoyan, H. (1994). *Fractals, Random Shapes and Point Fields. Methods of Geometrical Statistics.* John Wiley & Sons, Chichester.

Takens, F. (1981). Detecting strange attractors in turbulence. In: *Dynamical Systems and Turbulence, Warwick 1980. Proceedings of a Symposium held at the University of Warwick 1979/80. Lecture Notes in Mathematics 898,* 366–381. (Edited by D.A. Rand and L.-S. Young). Springer-Verlag, Berlin.

Takens, F. (1985). On the numerical determination on the dimension of an attractor. In: *Dynamical Systems and Bifurcations. Proceedings of a workshop held in Groningen, The Netherlands, April 16–20, 1984. Lecture Notes in Mathematics, 1125,* 99–106. (Edited by B.L.J. Braaksma, H.W. Broer and F. Takens). Springer-Verlag, Berlin.

Takens, F. (1993). Detecting nonlinearities in stationary time series. *Int. J. Bifurcation and Chaos* **3**(2), 241–256.

Taqqu, M.S. (1975). Weak convergence to fractional Brownian motion and to the Rosenblatt process. *Z. Wahrscheinlichkeitstheorie verw. Gebiete* **31**, 287–302.

Taqqu, M.S. (1977). Law of the iterated logarithm for sums of nonlinear functions of Gaussian variables that exhibit long range dependence. *Z. Wahrscheinlichkeitstheorie verw. Gebiete* **40**, 203–238.

Taqqu, M.S. (1978). A representation for self-similar processes. *Stochastic Processes & Appl.* **7**, 55–64.

Taqqu, M.S. (1979). Convergence of integrated processes of arbitrary Hermite rank. *Z. Wahrscheinlichkeitstheorie verw. Gebiete* **50**, 53–83.

Taqqu, M.S., Teverovsky, V. and Willinger, W. (1995). Estimators for long-range dependence: An empirical study. *Fractals* **3**(4), 785–798.

Taqqu, M.S. and Wolpert, R.L. (1983). Infinite variance self-similar processes subordinate to a Poisson measure. *Z. Wahrscheinlichkeitstheorie verw. Gebiete* **62**, 53–72.

Taylor, S.J. (1964). The exact Hausdorff measure of the sample path for planar Brownian motion. *Proc. Camb. Phil. Soc.* **60**, 253–258.

Taylor, S.J. (1986). The measure theory of random fractals. *Math. Proc. Camb. Phil. Soc.* **100**(3), 383–406.

Theiler, J. (1986). Spurious dimension from correlation algorithms applied to limited time-series data. *Physical Rev. A* **34**(3), 2427–2432.

Theiler, J. (1988). Lacunarity in a best estimator of fractal dimension. *Physics Letters A* **133**(4/5), 195–200.

Theiler, J. (1990). Statistical precision of dimension estimators. *Physical Rev. A* **41**(6), 3038–3051.

Theiler, J. (1991). Some comments on the correlation dimension of $1/f^\alpha$ noise. *Physics Letters A* **155**(8/9), 480–493.

Tricot, C. (1982). Two definitions of fractal dimension. *Math. Proc. Camb. Phil. Soc.* **91**, 57–74.

Tricot, C. (1995). *Curves and Fractal Dimension.* Springer-Verlag, New York.

Vere-Jones, D. (1995). Forecasting earthquakes and earthquake risk. *Int. J. Forecasting* **11**, 503–538.

Vere-Jones, D. (1999). On the fractal dimension of point patterns. *Adv. Appl. Prob.* **31**, 643–663.

Vere-Jones, D. (2000). Seismology - a statistical vignette. *J. Amer. Statist. Assoc.* **95**(451), 975–978.

Walters, P. (1982). *An Introduction to Ergodic Theory.* Springer-Verlag, New York.

Wang, Q. (1995). Fractal dimension estimates of global and local temperature data. *J. Appl. Meteorology* **34**(11), 2556–2564.

Wells, D. and Coppersmith, K.J. (1994). New empirical relationships among magnitude, rupture length, rupture width, rupture area, and surface displacement. *Bull. Seismol. Soc. Amer.* **84**(4), 974–1002.

Whitney, H. (1936). Differentiable manifolds. *Ann. Mathematics* **37**(3), 645–680.

Willinger, W. and Paxson, V. (1998). Where mathematics meets the internet. *Notices of the American Mathematical Society* **45**(8), 961–970.

Willinger, W., Taqqu, M.S., Leland, W.E. and Wilson, D.V. (1995). Self-similarity in high speed packet traffic: analysis and modeling of ethernet traffic measurements. *Statistical Sci.* **10**(1), 67–85.

Wolff, R.C.L. (1990). A note on the behaviour of the correlation integral in the presence of a time series. *Biometrika* **77**(4), 689–697.

Wyss, M., Aceves, R.L. and Park, S.K. (1997). Cannot earthquakes be predicted? (with response). *Science* **278**(5337), 487–490.

Young, L.-S. (1982). Dimension, entropy and Lyapunov exponents. *Ergod. Th. Dynam. Sys.* **2**, 109–124.

Index

Milton Keynes UK
Ingram Content Group UK Ltd.
UKHW040107071024
449327UK00019B/873